MANUAL OF SPECTROSCOPY

by

THEODORE A. CUTTING

CHEMICAL PUBLISHING CO., INC.
New York   N. Y.

Copyright, 1949
CHEMICAL PUBLISHING CO., INC.
New York N. Y.

PRINTED IN THE UNITED STATES OF AMERICA

CONTENTS

	PAGE
Preface	v

CHAPTER
- I. History and Theory of Spectroscopy 1
  - 1. Historical Review 1
  - 2. The Atom in Spectroscopy 3
- II. Light Sources 7
  - 1. The Electric Arc 7
  - 2. The Spark 15
- III. Spectroscopes 19
  - 1. Optical Systems 19
  - 2. Industrial Spectrographs 29
  - 3. Spectroscope Construction 38
- IV. Spectroscopic Analysis 58
  - 1. Qualitative Analysis 58
  - 2. Determinations 62
  - 3. Quantitative Analysis 71
- V. The Spectroscope in Mineralogy 88
  - 1. Tests Used in Mineral Identification 88
  - 2. Application of the Spectroscope in Mineral Classification 91
- VI. Characteristic Lines of the Elements 97
- VII. Wave-Length Table-Chart 186
  - Appendix 213
  - Conversion Table 213
  - Bibliography 214
  - Firms Selling Spectrographic Materials 215
  - Index 217

PREFACE

Too many chemists, mineral collectors, prospectors, and even assayers struggle with tedious chemical and uncertain flame and blowpipe tests when a spectroscope would give far more prompt results. With electricity universally available, and with present-day instruments and parts so low in cost, spectroscopic equipment should be in every school and laboratory.

This book has been written to assist those who wish to analyze ores, minerals, alloys, and inorganic chemicals, or wish to teach others to do so. In the qualitative analysis of such materials, there is no instrument so rapid and accurate as the spectroscope, although the analyst must remain within its limits of operation. This is also true of the quantitative analysis of these materials. Although speed comes only after some experience, one may very soon acquire the necessary technique for accurate determinations. The author has attempted to point out some of the short cuts to quick spectroscopic success. Direct methods of burning samples are shown; the key lines of each element have been selected, and a new chart-table has been prepared which shows both the spacing of spectral lines and their wave-lengths. An increasing number of schools and universities have courses in spectroscopy, and many industrial plants use spectrographic equipment to speed up the solution of their special problems of analysis.

The bulk and cost of spectroscopes tend to increase in geometric ratio with efficiency, and thus the price of a commercially built high-dispersion instrument is usually beyond the reach of the individual analyst. The parts necessary to make a powerful instrument are few and the essential construction is very simple. The writer believes that instructions for making effective instruments at costs so low as to be within the reach of all, will be appreciated. One section of the book has, therefore, been devoted to such instructions. Largely from his own experience, but also from the literature (see bibliography) a considerable amount of material has been arranged in what is hoped will be helpful form.

Acknowledgments. The writer wishes to express appreciation to those who have assisted in the preparation of the book: especially to his wife, Mary Elizabeth, without whose assistance the work would have been very difficult; to his sons, Windsor C. Cutting, M.D. and Cecil C. Cutting, M.D. for reading the entire manuscript, and for extensive criticisms and suggestions; to Henry Elliott, Ph.D. for reading portions of the manuscript and for constructive ideas; to Leslie Titus, A.B. for spectrum photographs and for reading the entire manuscript in the light of his special chemical and spectroscopic training and experience; to Frederick K. Vreeland, D.Sc. for special data; to his brother, James A. Cutting, M.D. and to the Applied Research Laboratories for photographs and literature; and to the San Jose State College for the use of special equipment.

CHAPTER I

HISTORY AND THEORY OF SPECTROSCOPY

1. Historical Review

Spectroscopy may be considered to have had its origin in the discovery by Newton, about 1670, that sunlight is composed of many colors. His experiment consisted in dispersing a beam of sunlight into its several colors by means of a glass prism. Huygens, a contemporary, developed the theory, contrary to the corpuscular idea of Newton, that light consisted of wave motions. Wave diagrams are still used to illustrate the properties of light.

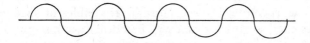

FIGURE 1
Light Waves

It was more than a hundred years before any further advance was made. Early in the 19th century Herschel revived interest in light phenomena by showing that the spectrum continued beyond the visible range. This he demonstrated by placing a thermometer in the infra-red portion of the spectrum. Ritter, in 1801, proved the extension of the spectrum into the ultra-violet by showing that silver chloride was darkened when placed there. Wollaston very soon afterward constructed the first real spectroscope with both prism and slit. With it he discovered the black lines of the solar spectrum. At about the same time, Young discovered the phenomenon of interference, which was later used for determining wavelengths.

Fraunhofer, in 1814, made the first diffraction gratings and used them for the first measurements of light waves. He also mapped the solar spectrum and even extended his observations to the spectra of the stars. Finally he recognized as identical the sodium doublet of the solar spectrum and the doublet of his laboratory flame.

Kirchhoff explained the black lines of the solar spectrum as caused by clouds of metallic vapors surrounding the white-hot orb of the sun. In 1861, he and Bunsen, the inventor of the Bunsen burner, became the originators of spectroscopic analysis by publishing the results of their investigations of the spectra of the elements. Three years later Maxwell contributed his electro-magnetic theory of light, and Balmer, about 1880, discovered a formula for computing the lines of spectral series.

Rowland, in 1881, ruled the first concave gratings and produced with them such superior spectra that accurate mapping of the solar spectrum became possible. His tables of spectral lines represented a great advance in practical spectroscopy. In 1901 Planck propounded his quantum theory of light which revived the corpuscular theory of Newton. Michelson, at about the same time, invented both the echelon and the interferometer, with which he measured the Paris meter bar in terms of the wave-length of the red cadmium line and so established a new and more permanent unit of measurement. In 1912 Bohr harmonized the earlier theories of electron movement with the quantum idea in his "Theory of Spectra and Atomic Constitution."

Since 1915, Meggars of the U. S. Bureau of Standards, has done much to stimulate the use of the spectrograph, especially of the grating type, for both qualitative and quantitative industrial analyses. About thirty of the elements were discovered by means of the spectroscope, and some of them can still be found only with difficulty, if at all, by any other means. Rubidium (red), thallium (green), cesium (blue), and indium (indigo) were all named from the colors of their most brilliant spectral lines.

2. The Atom in Spectroscopy

The Elements. Terrestrial forces, operating through eons of time, have formed about 1,000 mineral species, natural compounds, in the earth; man, in the last 200 years, has formed hundreds of thousands of other compounds; all of these are different combinations of 96 elements, each of which has its characteristic atom structure. Atoms, although inconceivably small and quite beyond the range of even the electron microscope, are made up of still smaller bodies, the most important of which are protons, electrons and neutrons.

Subatomic Particles. Electrons are particles of almost negligible mass, each carrying a single negative charge of electricity. Protons have about 1,840 times the mass of electrons and each one carries a single positive charge. Neutrons, with only slightly greater mass than protons, carry no charge. Positrons, which have the same mass as electrons and equal but opposite charges, are emitted by bombarded nuclei. The neutrino with electronic mass and no charge, and the negatron with a mass corresponding to that of the neutron and a negative charge, have also been reported.

Atomic Structure. It is the electron with which we have chiefly to deal in spectroscopy, since its motions produce light. Each atom consists of a nucleus containing protons, usually neutrons, and one or more electrons revolving around it like tiny planets about a sun. When the elements are arranged in their order of complexity, starting with hydrogen and ending with uranium and the new elements, neptunium, plutonium, americium and curium, their atoms are found to differ, each from its predecessor, by one proton, one electron, and a varying number of neutrons. Normally these electrons revolve about the nucleus in definite orbits, and, according to the theory of Bohr, in concentric shells with a definite number of electrons in each shell, as illustrated in Figure 2.

The periodic chart of the elements, Figure 2, is arranged spirally to show:

1. Atomic numbers in consecutive order

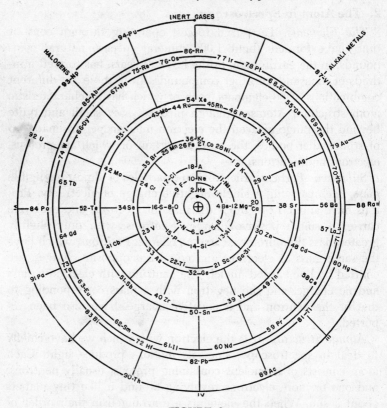

FIGURE 2
Periodic Chart of the Elements (Spirally Arranged)

2. The theoretical structure of each atom
3. The shell arrangements
4. The number of electrons in the outer shell
5. Chemical valences (Roman numerals)
6. Chemical relationships (Radial positions)

Light. In spectroscopy we are always analyzing light. The modern theory of light is complicated, involving emission by the atom of corpuscles or quanta of light in wave form. When an element is heated to incandescence, the motions of the electrons are accelerated and energy is absorbed. Moreover, the electrons leave their usual orbits and shells and move to higher levels more distant from the nucleus.

When the electrons return or fall again to their former levels, they give off the absorbed energy in the form of light. According to this theory a short fall produces long waves of red or infrared light, and a long fall short waves of violet or ultraviolet light. Upon excitation, the electrons of a given atom can assume only certain positions or orbits; consequently, in falling, they emit light of only certain colors or wave-lengths. Furthermore, since each element has a different number of electrons from the others, each emits a different assortment of colors when heated to incandescence.

Hydrogen, the simplest of the elements, when enclosed in a tube with a high-voltage current passed through it, will glow with a distinctive bluish light. When viewed through a spectroscope this light will be seen to consist, not of a single color, but of several; for the instrument sorts out the different colors of the light source and shows each as a brightly colored line. In the case of hydrogen the lines will be red, blue and violet, and they are said to comprise the hydrogen spectrum, or the visible portion of it, for there are also ultraviolet and infrared lines.

The speed of light has been accurately measured and is the same for all colors, but the number of waves or pulsations per second (frequency) is different for each color and for each shade of that color. The wave-length of the light of any spectral line is also different from that of any other, since it is the quotient of the velocity of light divided by the frequency of its wave.

The wave-lengths of the spectral lines forming the hydrogen spectrum fall into several mathematical series as though each group represented the light given off by electrons dropping from different heights to the same level, as shown in figure 2. In the larger group the short heavy line represents the electron fall causing the very brilliant red spectral line of hydrogen, and it is caused by the short fall of electrons from the third to the second level. The next two lines represent falls causing weaker blue

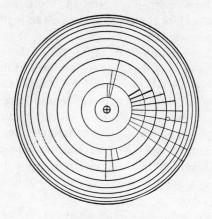

FIGURE 3
Hydrogen Atoms

lines; then come falls causing two violet and finally four ultraviolet ones. Other series to the right and left show electrons falling to the first and third levels. According to the theory, electrons may fall from level to level in short jumps, or even from the top to the bottom level in single long jumps. Mathematical formulae for these series have been developed by which, if a few spectral lines of an element are known, the wave-lengths of the other spectral lines of the series may be predicted.

Chapter II

LIGHT SOURCES

1. The Electric Arc

Light Sources. For spectroscopic analysis the rock or other sample must first be vaporized by heat and the incandescent vapor then viewed or photographed by means of a spectroscope or spectrograph, since it is only in the vaporized state that each element emits its distinctive colors or imparts them to the flame used for vaporizing it. The original heat source was the Bunsen flame, but its temperature is so low that it can vaporize few of the elements. To-day the electric arc, the spark and the discharge tube have taken its place. For different elements and for different types of samples different sources are required.

The electric arc is a continuous luminous discharge produced at the break in an electric circuit of low voltage. The spark is the discharge occurring at a gap in a high-voltage circuit. A potential of at least 40 or 50 volts is necessary to produce a sustained arc; from 10,000 to 25,000 volts is used in the spark. The discharge or Geissler tube is one filled with the rarefied gas to be analyzed; it is energized by passing a current of high potential through it.

Sources Required for Different Elements. Either the electric arc or the spark may be used to produce the known spectra of 65 of the metallic elements. The Geissler tube is employed for the 11 gases: argon, chlorine, fluorine, helium, hydrogen, krypton, neon, nitrogen, oxygen, radon and xenon and liquid bromine. None of the ten non-metallic or semi-metallic elements have very satisfactory visible arc lines: these are arsenic, antimony, boron, carbon, iodine, phosphorus, silicon, selenium, sulfur and tellurium. All of them, however, except iodine and sulfur have ultraviolet arc lines. The spectra of the latter two may be examined in either the discharge tube or in the spark.

Boron has no visible spectrum. A few of the lines and distinctive molecular bands of the *alkali elements* may be brought out in hydrogen or other gas flames, but the only satisfactory way to secure the complete spectrum of an element is to use electrical excitation. The major metals all show strong spectral lines, most of them many such lines.

The Arc. By attaining temperatures up to 8000° C, the arc vaporizes all rocks and chemicals and so brings out their full wealth of spectral lines. A potential of nearly 50 volts is necessary to sustain an electric arc, a 110-volt potential is better still. Most operators, in fact, choose 220 volts if available. Direct current is always preferred, but highly satisfactory results may be obtained with 110-volt alternating current. Only carbon electrodes will function with alternating current; with direct current rods of copper, silver, graphite, or other conductive material may be employed. The arc has been used as the chief source for producing spectra since the invention of the Gramm dynamo in 1876.

Portable Arc. Prospectors sometimes rig up a generator driven by the automobile engine and carry their spectroscopes with them into the field. Old 12-volt Dodge generators have been successfully rewound to deliver about four times the rated voltage and used in this manner. Regular generators for producing 5 amperes at 100 volts may be purchased or specially wound. The higher the voltage, the less the amperage required to work the arc. Since the number of watts is the product of the number of amperes multiplied by the number of volts, the following table, giving voltages commonly used in arcs, will show that it is customary to use about 1000 watts, except in the case of the high-voltage arc.

COMMON ARC CURRENTS

20 amperes at   50 volts or 1000 watts
10 amperes at 110 volts or 1100 watts
 5 amperes at 220 volts or 1100 watts
2.5 amperes at 2000 volts or 5000 watts

Resistor. A resistor is used in series with the arc to reduce the starting current to about 10 amperes, and to prevent drawing ex-

FIGURE 4
Resistor Types

cessive current and blowing fuses. One 1000-watt heating element of nichrome wire, or two 500-watt elements in parallel, will pass the proper amount of current at 110 volts. Forty feet of ordinary No. 22 *stone* wire will also make a satisfactory resistor, just enough of the wire being used to prevent its becoming red hot and burning out. A shorter length of No. 24 iron wire may be

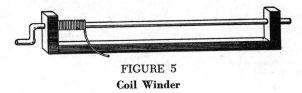

FIGURE 5
Coil Winder

employed with the same results. The wire is first coiled about a long iron rod the size of a lead pencil, which may well be mounted and bent at the end to form a handle for turning. Then the coil is slipped off the rod and wound upon a 10-inch length of asbestos-covered gas pipe or porcelain tube.

Horizontal Arc. The arc consists merely of two carbon rods or other electrodes held end to end, in any convenient manner, upon insulating supports. Carbons are preferably about ¼ inch in diam-

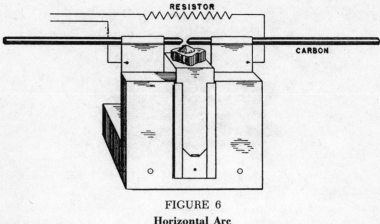

FIGURE 6
Horizontal Arc

eter, since larger electrodes draw more current than is necessary. The electrodes must be movable so that they may be pulled apart to provide the gap after making initial contact. If the slit in the spectroscope is horizontal, then the carbons should be arranged horizontally; if the slit is vertical, then the carbons are also vertical.

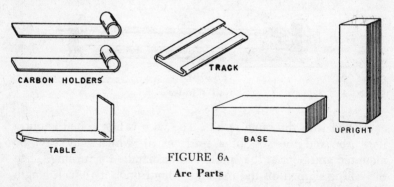

FIGURE 6A
Arc Parts

The following example will serve to illustrate a simple homemade arc. Nail together at right angles two small boards, each about three inches square, one to serve as a base, the other as an

Light Sources

upright. Next cut out two strips of any sheet metal an inch wide and three inches long for carbon holders. Curl one end of each strip around a spike slightly smaller than the carbons and then pull away the spike. Nail the strips to the upright about an inch apart with the curled ends an inch above the top. These will hold the carbons end to end in a horizontal position.

Prepare another piece of sheet metal, about 1¼ inches wide and 2 inches long, to serve as a track for holding the table strip. Fold over both of the long edges about one-eighth inch. Before flattening the folds completely slide in a double strip of tin and then press flat in a vise. Pull away the strip of tin and nail the track vertically to the middle of the upright on the opposite side from the carbon holders. Finally cut another strip about 1 inch wide and 4 inches long to slide smoothly within the folded track just prepared. Insert the sliding strip and bend it at right angles in the middle so that the horizontal part will form a table between the carbon holders and beneath the carbons upon which to place the sample to be tested. Be sure that the table slides clear of the carbon holders so as not to short out the carbons and so that the ore may be pushed up close beneath the arc, or a little into it. A nib should be bent out at the bottom of the strip to facilitate pushing it up and down after the top becomes hot from the arc.

Wiring the Arc. Separate the twisted wires at the end of a heavy drop cord. Connect one terminal directly to one of the carbons or to its holder and connect the other to one terminal of the resistor. Finally connect the opposite terminal of the resistor to the second carbon holder. This provides a circuit for the current to come down one of the twisted wires of the drop cord, pass through the resistor, go to one carbon, jump the gap to the other, and finally go back up the second wire of the drop cord to complete the circuit. The resistor must always be included in the circuit to prevent shorting and blowing fuses.

The remote ends of the carbons may be taped to prevent shocks during manipulation, although if one remembers to touch but one carbon at a time, and the floor is dry, this is unnecessary. There is little danger from a 110-volt circuit, but linoleum should be used on cement or damp floors to prevent the uncomfortable shocks of a ground circuit. With the use of 220 volts twice the precautions are necessary against shocks and against shorts as well. One should look at the arc flame only through smoked or cobalt-blue glass to avoid injury to the eyes from the ultraviolet radiations.

High-voltage Arc. Excellent results are reported by many operators who have used high-voltage alternating-current arcs. In these, the voltage is stepped up by a heavy transformer to about 2,000 volts, and a current of 2½ amperes is used. There is no condenser in the circuit. In the quantitative analysis of metallurgical products it is claimed to be superior to the direct-current arc.

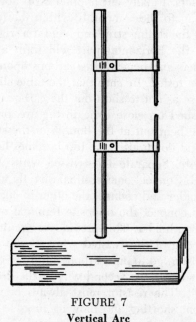

FIGURE 7
Vertical Arc

Vertical Arc. For vertical slits the carbons are clamped, end to end, one above the other on insulating supports. A little cup is drilled in the top end of the lower carbon to hold the pulverized sample. Such an arrangement results from turning the horizontal arc previously described on its side and removing the ore table; two metal strips with curled carbon-holding tips may also be fastened, one above the other, by bolts on an upright length of broomstick.

Manipulations. Volcanic rocks, shales, cherts, and many ores can be analyzed by burning corners of inch-high fragments in the arc. The specimen should be turned over frequently and new corners burned, otherwise some of the elements present in small proportions will become vaporized and pass off before being observed. It is also well to pulverize some of the ore and to add a bit of it frequently with the spatula to the top of the fragment being burned. In this way, line-brilliance may be maintained as long as desired. Each part of a composite rock may be burned in succession to determine the composition of each crystal, small vein, or differently colored portion.

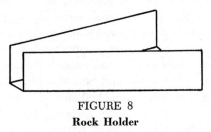

FIGURE 8
Rock Holder

Pieces of sheet metal, two inches square or of trapezoidal shape, bent up at opposite edges to form little flat-bottomed troughs, will afford very convenient clamps to hold the rock fragments or quartz tables while burning. If the trough is made wider at one end than at the other, the fragments may be wedged in tightly.

Sometimes beads will form on the carbon points and make reestablishment of the arc difficult. A third carbon rolled over the tips will often start up the arc again. At other times the point will have to be ground or broken away and a new start made. Before beginning a new analysis the carbon tips will usually have to be carefully cleaned by sandpapering or grinding and wiping with a cloth. If spectral lines still remain from the previous analysis, a few seconds of burning will often clear the arc. When the test shows no held-over lines the new sample may be inserted. Traces of calcium and iron may be present in the carbons themselves, but *National* carbons (specially made for this work) are very pure. Ordinary projector carbons are also satisfactory. If one wishes to observe the flame of the arc he may do so by using a lens to project an image of the arc upon a white cardboard set up behind the carbons.

Arc Limitations. The hydrogen and carbon of organic compounds cannot be determined in the carbon arc, nor the oxygen content of rocks. Neither can the nitrogen and phosphorus content of soils be ascertained without resort to the Geissler tube. The metallic constituents of organic substances can be determined most easily by ashing the compounds and then analyzing the ash. Mineral waters and solutions of all kinds are usually evaporated and the residue analyzed, although they may be added drop by drop to powdered silica, and examined directly.

Ores rich in chromium, iron, copper and similar metals will often melt down into masses too conductive and too large for vaporization, or will bubble and boil and blur the spectrum. In such cases, if the sample is removed from the arc, enough of the molten sample may cling to the carbon points to continue the spectrum. At other times it will be necessary to break off a smaller fragment of the ore and pulverize it, together with a similar amount of pure quartz or carbon in a mortar. The diluted powder, burned on a little flat piece of quartz rock, will then give a satisfactory spectrum, since silicon yields no strong visible lines. Metal filings, if mixed with powdered quartz, may be similarly burned. Optionally, too-conductive material may be handled by taking a very small sample, about the size of a grain of wheat, and burning it on a quartz block without the addition of silica.

Crystals that fracture in the intense heat of the arc, and snap and fly in all directions must be finely crushed before burning.

Light Sources

Liquids may be dropped upon powdered quartz or upon aluminum oxide, since alumina, also, has no troublesome visible lines. The beginner should realize that he must have vaporizing material in the arc or there will be no spectrum. Replenish the material freely for some elements disappear quickly.

Nitrates, sulfates, chlorates and other similar chemicals will often fluff up and recoil from the arc without giving a satisfactory spectrum, or form little molten balls that roll away. The solution for all such troubles is to pulverize the chemical and a similar quantity of quartz or aluminum oxide together and burn the mixture upon a fragment of quartz or chalcedony of not too crystalline character. Since veins of milky quartz are common in most rock masses, such fragments are usually easily obtainable.

Bits of porcelain tubes or blocks such as are employed for electrical insulation may be used in place of quartz although they will usually contain some iron, titanium, or calcium, with aluminum.

2. The Spark

Characteristics of Spark Spectra. Arc and spark spectra of the same element usually contain enough strong lines in common to make their relationship easily recognizable, but they also have very noticeable differences. In general the arc lines will be stronger than the spark lines—the sodium doublet, for instance, is ten times as strong in arc as in spark spectra; occasionally, however, a spark line will be stronger than the corresponding arc line. Certain arc lines may even be missing from the spark spectrum or vice versa.

Ionization. When an element is subjected to a high-potential current, an electron may be driven entirely out of its atomic system. The atom will then have the same number of electrons as the next lower element in the atomic series and the two spectra will have resemblances. Excitation by currents of still higher potential may cause the loss of more electrons and produce resemblances of spectra to those of atoms still lower in the atomic series.

Transformer. For spark spectra a transformer stepping up the 110-volt current to 15 or 20 thousand volts is necessary. Such a transformer should have a closed core of laminated iron with a cross-section of 2½ square inches. The primary winding may have 300 turns of No. 16 copper wire and the secondary winding 41,000 turns of No. 35 enameled copper wire, with strips of paraffined paper between the layers for protective insulation in both windings.

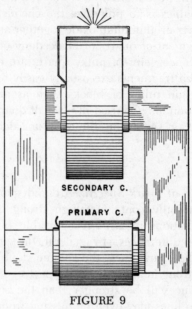

FIGURE 9
Transformer

Of course, the high-voltage apparatus necessary for spark spectra is dangerous, especially the condenser and the spark gap. Although shocks from them may not be fatal, they are exceedingly painful and in an altogether different class from the tingling shocks given by the 110-volt circuit. A good arrangement is to enclose everything connected with the high-voltage apparatus in a cage with a door which opens the switch when it opens.

Light Sources

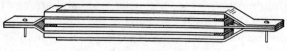

FIGURE 10
Condenser (Partially Assembled)

Condenser. A condenser made of a dozen 8 in. x 10 in. glass plates, each consisting of double-strength window glass, alternating with thin sheets of metal, is connected in parallel with the gap to intensify the spark. Moreover a single-layer, air-core induction coil about two inches in diameter and having 100 turns of No. 20 insulated copper wire must be used in series with the spark gap to eliminate the spectral lines of elements in the air that would otherwise fill the spectrum. A grounded screening cage or large condensers across the 110-volt leads must be used if radio interference is to be avoided. A two-millimeter safety gap should be placed across the alternating current leads to prevent dangerous kick-backs into the power line. Finally, a resistor should be provided in the primary circuit; this will be similar to that used in the arc circuit to prevent the flow of more than one kilowatt. Wiring is shown in the diagram.

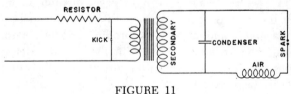

FIGURE 11
Spark Circuit

Gas Tube. An optional method for eliminating the oxygen and nitrogen lines of the air is to produce the spark in a closed tube containing illuminating gas since hydrogen produces only a few strong lines at atmospheric pressure. Such a tube may be made as follows: Drive the air from a short glass tube by a few seconds' flow of gas from the main and then quickly cork the ends. Perforate each cork with a metal wire, and arrange the two wires to form a short gap at the center of the tube. Practically all of the air must be forced out or an explosive mixture will be formed that will blow out the corks. When the wires are connected to the high-voltage terminals, the spectrum at the gap will be a combination of a few hydrogen lines from the gas and those of the electrodes.

Spark spectra are frequently preferred for the analysis of smelted metals and their alloys. Nonconductive substances may be mixed with powdered carbon and pressed into pellets for use as electrodes. The spectrum of elements in solution may be obtained by allowing the solution to drip slowly from an inverted cone of filter paper to a receptacle beneath. A conducting wire is inserted into the liquid above and another into the receptacle below. A spark is then passed between the drop at the tip of the cone and the liquid surface below.

Tesla Coils may be used as a spectrum source; however, in spite of the tremendous voltage, the lines are usually weak and unsatisfactory, in the experience of the writer.

Discharge Tubes. Rarefied gases (such as oxygen and nitrogen), when enclosed in Geissler tubes, will glow and yield distinctive spectra when a high-voltage current is passed through them. The spectra of even the inert gases, helium, neon, argon, etc., may be secured in this manner. Likewise the spectra of sulfur, iodine and selenium may be produced in evacuated tubes by covering the lower electrode with them in powder form. In such tubes the degree of evacuation affects both the number and intensity of the spectral lines.

Chapter III

SPECTROSCOPES

1. Optical Systems

Function. The spectroscope is a very simple instrument; nevertheless, it is one of the most useful developed by science. By sorting light waves according to their lengths it can show which elements are present in either a grain of sand or a distant star. It enables one to read off the elements present in a rock or ore with almost the ease and speed of reading from a printed list and, since the strength of spectral lines is proportional to the number of atoms excited, it also makes possible valuable quantitative estimates.

Types. There are several types of spectroscopic instruments employed for separating light waves into their different colors or wave-lengths. Prism and grating spectroscopes are used for practical analysis, and echelons and interferometers for the exact measurement of light waves in research. Prisms give maximum brilliance and increasing dispersion toward the violet end of the spectrum; gratings give ample brilliance and equal dispersion over the entire spectrum.

Gratings. The diffraction grating is merely a flat or curved surface of glass or metal ruled with very fine parallel lines, usually from 15,000 to 30,000 per inch. Johns Hopkins University, Chicago University, and the California Institute of Technology have machines of the requisite precision for this work. Light passing through transparent gratings or reflected from metal ones is fanned out into its separate colors.

One may secure a spectroscopic effect by holding a finely ruled transparent grating near the eye and looking through it at an electric arc a few feet distant. Brilliantly colored spectra will appear on opposite sides of the arc and spaced some distance from it. If a bit of table salt or some other sodium compound is introduced into the arc, a highly intensified yellow spot caused by the sodium will appear in the yellow spectrum. Careful inspection will show that lesser spots also appear in the red, green and blue sections—for no element produces a spectrum so simple as to consist of but a single wave-length of light.

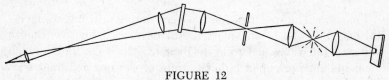

FIGURE 12
Optical System of the Spectroscope

Slits. If the light from the arc is made to pass through a narrow slit, as between the edges of two razor blades, or a needle scratch on silvered glass, the colored spots become spectral lines, and the arrangement of slit and grating is a spectroscope. When the slit is made narrow enough, the single yellow line will be resolved into two, the sodium doublet. If the salt in the arc is replaced with a copper compound, the yellow lines will fade and three brilliant green ones in the adjoining part of the spectrum will appear. Zinc in the arc will produce three bright blue lines of still different wave-lengths. The slit must be very narrow and clean-cut to prevent the blurring or blending of spectral lines; dust and rough edges make troublesome shadows. Spectral lines are images of the slit colored by the gases in the heat source.

Prisms. Glass or other transparent prisms will also disperse light, sending the red waves to one end of the spectrum and violet ones to the other. In *direct-vision* spectroscopes the light passes through a train of prisms made alternately of crown and flint glass so as to disperse the light into its separate wave-lengths and to refract it so that it will emerge in about the same direction that it entered.

Collimator. A lens, usually of from 6 to 10 inch focal length, placed between slit and prism at a distance from the slit equal to its focal length, converts the light which falls on it into a beam of parallel rays and passes it on to the prism. This is the collimating lens. The focal length of a lens may be roughly determined by placing the lens in the sun and measuring the distance between lens and *burning spot;* the lens may also be taken to the far side of a room and the distance between the lens and the sharp image of the window (as cast upon a piece of white paper) measured.

Condensing Lens. A lens (usually of short focus) placed between the arc and slit, to concentrate the light upon the slit, is called a condenser. With prism instruments it is placed a little closer to the arc than to the slit so as to throw a somewhat enlarged image of the arc upon the slit. The total distance between arc and slit will be about 4 times the focal length of the lens. The effect is to place the arc right in the slit and to increase greatly the brilliance of the spectral lines and the blackness of the field.

Telescope. When magnification of the spectrum is desired, and it usually will be desired for analytical work, a telescope consisting of an objective lens and an ocular is used. Telescopes ordinarily view only distant objects, but since a properly placed collimating lens converts the light entering the slit into a beam of parallel rays as though coming from a great distance, the telescope can focus upon the slit and give a sharp image of it even though it is only a foot or two away. Telescope objectives usually have a focal length of from 6 to 10 inches and are placed near the prism on the opposite side from the collimator.

Ocular. The ocular usually has a focal length of about an inch and is placed so that its distance from the objective is a little more than the focal length of the objective. The telescope increases both the length of the spectral lines and the distance between them. A combination of two spaced lenses instead of one is often used for the ocular to increase the field of vision and to decrease chromatic aberration.

Instrument Similarities. In the prism spectroscope, the light from the arc passes in succession through condenser, slit, collimator, prism, telescope objective and ocular to the eye. The arrangement in the magnifying diffraction-grating spectroscope is exactly the same except that a grating is substituted for the prism. In direct-vision spectroscopes, both prism and grating, the telescope is omitted and so there is no inversion of the spectrum by lenses.

Lenses. Curved spectral lines are due to the faulty centering of lenses or to the nature of the prisms. Convex lenses, since they are thicker in the middle than at the edge, give color effects similar to those produced by prisms, causing what is called chromatic aberration. By combining lenses of flint and crown glass having different indices of refraction, these color defects are eliminated and the spectrum improved. Spherical aberration, still another type of lens distortion, may be corrected almost completely by compound lenses in the same manner. Such corrected or *achromatic* lenses give clearer spectra, but their cost is sometimes prohibitive. Simple lenses will often give surprisingly good results.

Scales. The best position for the scale is between the telescope objective and the ocular, at about an inch distance from the latter, for it is then alongside the spectrum so as to be viewed simultaneously with it and in sharp focus on the ocular. Prisms are of various shapes, the equilateral type being most common in moderately priced instruments. More powerful types of instruments sometimes employ two prisms in succession to double the dispersion, or reflect the light and pass it back through the same prism a second time. Still others use multiple-face prisms with some of the faces serving as reflectors, or mirrors, to reflect the light back and forth several times.

Spectrographs. For photographing the spectrum it is especially important that the interior of the instrument is painted a flat black to prevent reflections. The case must be lightproof except for the slit, and baffles are often used to trap stray light. A pan-

Spectroscopes

chromatic film is placed directly alongside the scale, and the ocular is removed or swung out of the way. Less than half of the spectrum is in the visible range, the remainder being in the ultraviolet and in the infrared. To photograph the rich ultraviolet spectrum, quartz prisms and lenses must be used since glass is opaque to light in this region. For the infrared, prisms of crystal salt or of sylvanite and special films sensitive to this type of light must be used.

Replicas. Although, in general, gratings give slightly less brilliance than prisms, good ones will give all the light necessary for both visual inspection and photographic recording. There is less scattered light in the grating spectrograph and, therefore, the field is darker. Replica gratings are made by pouring a solution of guncotton in amyl acetate on an original grating, allowing it to dry thoroughly and then floating it off in water. If carefully made and mounted on glass, the first fifty or so replicas are excellent, but if they are stretched in handling, or too many are taken, they give weak lines and troublesome phantoms—false fuzzy lines that do not belong to the spectrum. The lines of a 15,000-line grating are too fine to be seen except with a powerful microscope.

Concave Gratings. In many respects the most satisfactory spectroscopes made are those employing concave gratings. They magnify the spectrum without the need of auxiliary lenses and may often be used even without condenser lenses. An eyepiece is used, but for photographic recording this, too, is eliminated. The light from the arc passes directly through the slit to the grating and thence is reflected back to the ocular. The carefully-ground concave grating is silvered, or better still, aluminized in a vacuum, so as to reflect the light as completely as possible.

From the following table showing the dispersion of some of the concave gratings in common use it will be seen that doubling the focal length of the grating also doubles the length of the scale; that the spectrum of the second order is twice as long as the first; and that doubling the number of rulings per inch roughly doubles the dispersion. One millimeter on the scale of a grating with a ten-foot radius with about 24,000 lines per inch would represent one Ångström. In other words, the dispersion of a concave grating is proportional to the focal length and roughly proportional to the number of lines per inch.

TABLE 1

Dispersion of Concave Gratings

Focal length meters	Lines per inch	First order Å per millimeter	Second Order Å per millimeter
1.0	15,000	15.0	7.5
1.5	24,000	7.0	3.5
2.0	24,000	5.2	2.6
3.0	15,000	5.0	2.5
2.0	36,000	3.4	1.7
10.0	30,000	0.8	0.4

Dispersion. Spectroscopes vary greatly in size and in resolution. In small instruments the spectral lines often become crowded or even blended; in more powerful ones, the same lines will be separate and distinct. What may appear as a broad single line in a small spectroscope will often be resolved in a larger one to groups of several lines. For pure chemicals, simple compounds and ores, small instruments without magnification are often suitable, but for the complete analysis of complicated specimens, such as ordinary rocks and minerals, greater power is necessary. When mere traces of the rare earth elements are to be detected, spectrometers with scales or spectrographs which photograph the spectra are necessary.

Resolution. Dispersion relates to the fanning out of the lines; resolution to the separation of lines, which are close together, so that they are sharp and distinct; this depends on the quality of certain parts of the spectroscope. An instrument of high dispersion may give such fuzzy lines that there is less resolution than with a smaller instrument. Good small instruments will often resolve the sodium doublet with its component lines 6 angstroms apart, but magnification is necessary to resolve the lines of the chromium triplet, which have an average separation of less than two angstroms.

Large Instruments. Some spectrographs used in research separate lines of wavelengths differing by less than 1/1000 Å, but such accuracy is not required in analytical work. With the even dispersion of the grating one can soon learn to estimate the separation of lines either in the visible spectrum or in the spectrogram, since two lines differing by 6 Å will be the same distance apart as two violet, green, orange or red ones. With the prism, of course, this is not true, since much greater dispersion is shown in the violet than in the red. The U. S. Bureau of Standards usually prefers a quartz spectrograph in the region from 2100 to 2500 angstroms, and a diffraction grating instrument from 2500 to 9500 angstroms. The instruments are often twenty or thirty feet long. The dispersion of a 60° prism is equivalent to the dispersion of a 2000-line grating in the red portion of the spectrum and that of a 14,000-line grating in the violet. The quartz prism has less dispersion than a 10,000-line grating near the violet, but as much as a 90,000-line grating at 2000 Å.

Horizontal Mountings. The prism instruments found in schools and many smaller laboratories are usually arranged with their optical parts on a horizontal plane. The prism is clamped upon a central support; the telescope is trained upon one of its faces, the slit and collimator upon another, and light is directed through a transparent scale so as to be reflected from one of the prism faces into the telescope. Telescope, collimator and scale are in separate tubes pointing toward the central prism. The scale in such instruments is often arbitrary with readings in millimeters instead of in Ångström units. One of its lines is customarily marked with the symbol "Na," and before attempting to use the instrument, the prism must be adjusted so that the sodium doublet registers with it.

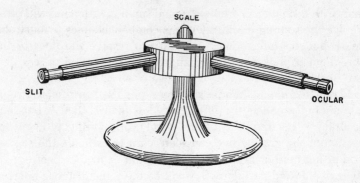

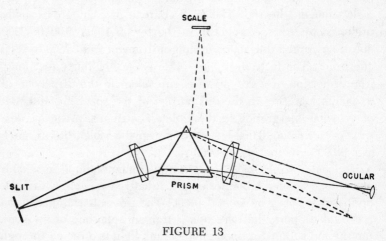

FIGURE 13
Horizontally Mounted Spectroscope

Graph. The conversion of the arbitrary readings of the instrument to Ångström units may be accomplished by means of a graph in which the numerals at the left represent hundreds of angstroms and those at the bottom the arbitrary figures of the scale. Dip the tip of one carbon in lithium chloride and carefully note the reading of the red line on the scale. We will suppose that it reads 0.2; then follow that line up until it crosses horizontal line 67 hundred, since the wave-length of the red lithium line is given in the tables as 6707. Place a dot where the two lines cross. We will also suppose that the orange lithium line reads 1.3

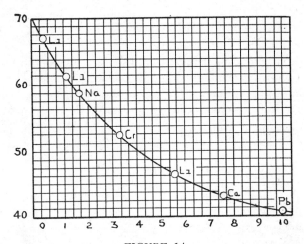

FIGURE 14
Scale Conversion Graph

on the scale; follow this line up until it crosses the horizontal graph line 61 hundred, since the wave-length of the orange lithium line as given in the tables is approximately 6100. Place another dot at this crossing. Similarly plot the sodium doublet 5900, the blue lithium line 4600, the chromium triplet 5200, the strong calcium line 4225 and the lead line 4050.

Finally connect the plotted points by a curve and test its accuracy by tracing lead, zinc and copper lines of known wavelength. When the curve is once established, the wave-length of any spectral line may be determined by finding the horizontal line that crosses the vertical line indicated by the scale and also crosses the curve at the same point. For instance, if a bright line appears in the instrument at scale reading 5.5 the wave-length, as determined by going up line 5.5 to the curve and then following the horizontal line to the left, is 46. From the table it will be seen that the two strong lines near 4600 are the lithium line of intensity 800 and the strontium line of intensity 1000.* To complete the identification one must look for other lines to see which is present in the arc, lithium or strontium; otherwise the lines must be measured more precisely if the dispersion of the instrument permits.

Another method of using an arbitrary scale is merely to make a table showing scale readings in one column and the corresponding wave-lengths in the other. Presupposing the same instrument used in making the graph the table would read:

TABLE 2

Conversion Values

Scale Reading	Angstroms	Scale Reading	Angstroms
0.0	6750 Fe	4.0	5000 Ti
0.5	6700 Li	4.5	4903 Fe
1.0	6154 Na	5.0	4704 Cu
1.5	5950 Si	5.5	4600 Li
2.0	5700 Cu	6.0	4554 Ba
2.5	5500 Fe	6.5	4401 Ni
3.0	5300 Co	7.0	4315 Fe
3.5	5202 Fe	7.5	4226 Ca

By filling in the spaces as further known lines are recorded, and by extending it to further fractions, one may estimate by interpolation any spectral line observed in the instrument.

Power of lenses. The magnifying power of a lens held close to the eye may be defined as the ratio of the size of the image as viewed through the lens to that of the object as seen without the lens at a distance of ten inches. The power of some of the lenses often used in spectroscopes is given in the following table:

---

* Line intensities range from 1 (very weak) to 9000 (very brilliant). Intensity is a very important factor in the identification of a line, as will be explained in the chapter on analysis.

TABLE 3

Magnification Power of Lenses

Focal Length, inches	Magnification
½	21
1	11
2	6
4	3.5
5	3
8	2.25
10	2
20	1.5

The magnifying power of a telescope is the ratio of the focal length of the objective to that of the eye-lens. The power therefore of the usual telescope used in the spectroscope is 8 or 10 to 1.

The spherical aberration of lenses may be decreased either by using two of lower power, instead of one strong lens, or by masking off the rims, or both. Of two lenses that may be used for the same purpose choose the weaker one, and avoid all unnecessary lenses.

2. Industrial Spectrographs

Technical Instruments. The finest spectrographs made to-day are those using precision-ground and polished quartz prisms and lenses, and those using precision-ground and precision-ruled original diffraction gratings. Both give such excellent quantitative results that they have supplanted the chemical analysis of alloys in determining their desired or undesired constituents up to about 5% of content. For the quantitative analysis of carbon, sulfur and sometimes phosphorus, however, chemical methods are still preferred since the spectral lines of these elements are unsuitable for such determinations.

Use of the spectrograph is especially advantageous in testing steels and other alloys while they are still in the melting pot, because of the speed with which this can be done. High-grade steels

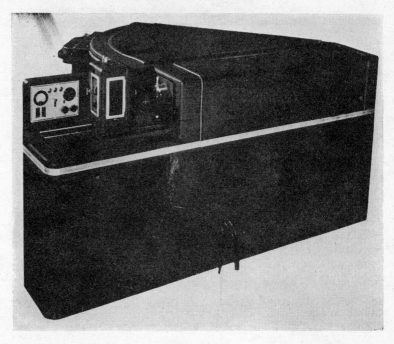

FIGURE 15
Applied Research Laboratories Spectrograph

and many other special alloys are easily impaired by the presence of very small quantities of impurity elements, or by improper proportions of those necessary to give the desired qualities. Pouring of the melt must await quantitative determinations, and time, fuel and labor are all saved by the use of speedy spectroscopic tests.

The spectral lines used in such analyses are almost always persistent ones, but are seldom the strongest lines of the element. Samples are Mo-2816, Cu-3274, Sn-3262, V-4379, Mn-2933 and Fe-4325.

Different operators report that they can determine with sufficient accuracy for the most exacting metallurgical requirements

Spectroscopes

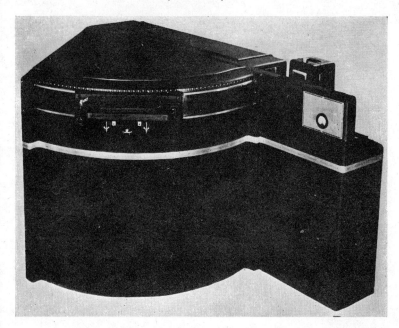

FIGURE 15A
Applied Research Laboratories Spectrograph (Another View)

amounts ranging from 0.001 per cent up to 2 or 3 per cent of most metals. Some elements such as silica, aluminum and iron may often be determined even up to a content of 10 or 15 per cent.

Graphite electrodes are usually preferred, and exposures average about 4 seconds, during which time the sample of about 25 mg is completely volatilized. For testing alloys the spark or the alternating-current arc at about 5000 volts is usually preferred. A condensing lens is used for focusing the image of the arc upon the slit of prism instruments, or upon the grating of concave grating instruments. In photographing the spectrum it is customary to include the blue, violet and much of the ultraviolet portion of the spectrum.

FIGURE 16
Quantometer, A R. L. (Dietert)

The Quantometer. The last word in spectrometric determination is an instrument called a *Quantometer* which automatically and quantitatively gives the contents of the sample in a direct reading. Analysis for as many as a dozen different elements takes little more time than the few seconds required to burn the sample.

No photography is required and thus all the uncertainties of film densities are eliminated. A single spectral line of each element is used. The light of the selected line, instead of striking a photographic plate, passes into an electron multiplier phototube, which passes more or less electric current according to the amount of light received. The electric current then operates other automatic devices and finally passes on the results to a recording console, where the percentages of all the elements appear on a panel. Any lines between 2000 Å and 6000 Å may be chosen for intensity measurement. Interfering lines as close as one angstrom may be avoided.

In view of the greater speed and accuracy attainable, the cost of the unit above that of ordinary spectrographic equipment is soon repaid, for it is a costly thing to hold up the pouring of melts until analytical reports can be made.

Infrared Spectroscope. Crude oil is broken up and fractionated into different grades of fuels and lubricants. The infrared spectroscope is used in testing such products as high-octane aviation gasolene and toluene for synthetic rubber. It gives greater speed in analysis and greater accuracy with some hydrocarbons than chemical and other physical methods.

The molecules of petroleum compounds contain atom masses which, when set into vibration by infrared radiations, absorb those vibrations with frequencies equal to their own. The absorption of the radiation from glowing oxide rods, as of cerium or thorium, is different for each compound, since each has different atom combinations, and therefore absorbs different wavelengths of the energy.

The infrared waves are dispersed by prisms of potassium bromide or sodium chloride and are made to fall in turn upon a thermocouple whose voltage output varies with the radiation received. A vacuum tube is used to amplify the thermocouple output which is then photographically recorded. The apparatus may be made sufficiently sensitive to detect potentials of less than one billionth of a volt. Both qualitative and quantitative analyses of hydrocarbons may be made from the spectrograms.

The mass spectrograph is also used in the oil industry. With it, operators have analyzed a mixture of a dozen hydrocarbons within three hours. In this process, the gas to be analyzed is introduced into a highly evacuated chamber and there bombarded by electrons from a heated filament, which ionize the gas. The speed of the ions is then accelerated by electrically charged plates and passed into a strong magnetic field. Here they are turned from straight paths into curved ones by the magnetic attraction, and since all the ions have the same initial charge and speed, the curves taken will depend upon their mass. This sorts out the heavy ions from the light ones and makes it possible to collect those of the same kind and to measure the current which they produce. By varying the voltage of acceleration, ions of different mass may be brought in turn to the collector and the varying current recorded as a mass spectrum.

The mass spectrograph can analyze more complex mixtures than the infrared spectrograph, but each instrument can analyze certain hydrocarbons more efficiently than the other.

Accessories

Film Holders. Spectrographs are provided with film holders, built into the instrument but usually removable so that they may be taken to a dark-room for the development of films. Means for lateral movement after each exposure are often provided so as to make possible several spectrograms side by side on the same strip of film.

Multisource Unit. Different samples often require different light sources for the most dependable quantitative results. Although the direct-current, low-voltage arc is the most versatile light source, since it can be used for either metals or non-metals, it is less convenient than the high-voltage spark for metals and alloys. The alternating-current arc is especially suitable for the analysis of metals and non-metallic salts, but it is difficult to control exactly.

Manufacturers of spectrographic equipment have designed compact units combining condensers, resistors, inductances, voltmeters, ammeters, phase controls, rectifiers, and timers in a single housing so that any type of light source desired can be used and so controlled as to give uniform results. One firm provides an oscillograph so that even the wave form of the discharge can be studied.

Comparator. The comparator is another valuable accessory. It is a device having lenses and mirrors for projecting the magnified image of the negative of the spectrum of the sample alongside a master scale which has numbers and labels indicating the

FIGURE 17
Multisource Unit, A. R. L.

important spectral lines of the elements. An iron spectrum may also be projected alongside that of the sample for the exact spotting of lines.

FIGURE 18
Projection Comparator-Densitometer

Densitometer. The densitometer is an instrument for measuring the density of the spectral line as photographed on the film. The light passing through the film at any spectral line falls upon a photocell, the output of which is amplified and then made to activate a galvanometer. The density as well as the identity of the line may be simultaneously determined.

Other Accessories. Other accessories often found in a well-

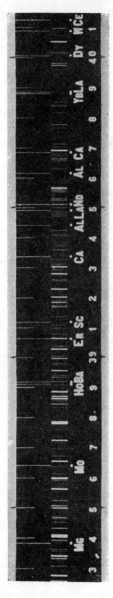

FIGURE 19
Comparator Scale

At the top are short lines 10 Å apart with intermingled longer key lines of the elements. In the middle is shown the spectrum of iron from 3825 Å to 4015 Å. Then come the symbols of the elements whose key lines are in this region, and at the bottom the scale numbers from 3825 Å to 4015 Å.

equipped spectrographic laboratory are molds and presses for compressing samples into electrodes, hoods for removing fumes from the spark or arc, light sources for absorption spectra, rectifiers for converting 8 or 10 amperes of alternating current to 220-volt direct current, transformers for the high-voltage arc and spark, inductance units for the elimination of air lines in the spark, arc and spark stands, and developing outfits for the films.

Procedure. By means of the accessories described, a film, exposed, timed and developed under carefully controlled conditions, is inserted in the comparator, whose labeled scale lines coincide with certain spectral lines of the film and so show what elements are present. Here it may also be compared with standard films for quantitative estimates if desired. Usually, however, it is next inserted in the densitometer for determining the density of key spectral lines and so the percentage of each element contained in the sample. In some cases the comparator and the densitometer are combined in one instrument.

Spectroscope Prices. Small box spectroscopes with just enough power to separate the lines of the sodium doublet may be purchased for $2.50. Metal instruments of about the same power but giving longer spectral lines cost from $10 to $20. Instruments that magnify the spectrum by means of lenses and contain some sort of scale are priced from $50 to $200. Spectrographs for photographing the spectrum, both visible and ultraviolet, may cost anywhere from $1,500.00 to $4,000.00, whether using concave gratings or quartz prisms and lenses.

3. Spectroscope Construction

In the following pages directions are given for constructing several different types of instruments, some simple, some more complicated. No attempt has been made to follow conventional designs, since these would be difficult to copy in most home shops. The aim has been to show how to build simple, efficient, inexpensive instruments with ordinary tools.

Spectroscopes

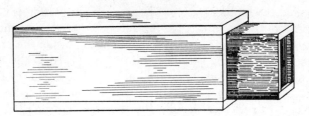

FIGURE 20
Plane Grating Spectroscope (Box Type)

PLANE GRATING SPECTROSCOPE. For a simple box-type spectroscope plane gratings ruled with either 15,000 or 25,000 lines per inch will have sufficient dispersion without magnification to show the characteristic spectra of the different common metals, but those of less than 10,000 are about as ineffective as single prisms. The case for a small spectroscope may consist of four wooden strips nailed together to make a tube two inches square and five inches long. Before assembling, grooves are sawed ⅛ inch from one end of each strip to hold a heavily silvered glass square with a horizontal needle scratch through the silvering for a slit. Insert the mirror-slit and nail the tube together. Optional slits are described in connection with some of the other types that follow.

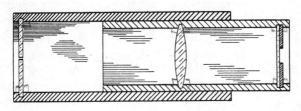

FIGURE 20A
Plane Grating Spectroscope Longitudinal Section

Next, prepare a shorter tube sufficiently smaller to slide smoothly within the first. Two sets of similar grooves are sawed in the inner tube—one set to hold the 4-inch focus lens, the second to hold the replica grating. The transparent grating is placed at one end of the spectroscope, the slit at the other, and the collimating lens in between as illustrated. The rulings of the grating must be parallel to the slit. The collimator is fixed in the drawtube an inch behind the grating; and when the inner tube is adjusted so that the lens is at a distance from the slit approximately equal to its focal length, the spectral lines will be sharp and clear.

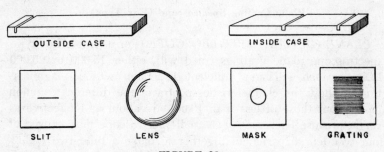

FIGURE 20B
Parts for the Box Spectroscope

One does not look horizontally through the tube, but down at an angle of about 30°. If desired, one can, by placing a 45° prism between collimator and grating, straighten out the beam of light, but since the spectrum is not improved, the insertion of extra prism is not worth the trouble. A mask with a peephole may be inserted in the grooves next to the grating to exclude light and prevent reflections; of course, as with all spectroscopes, the interior must be painted black. Such an instrument will show a multitude of Fraunhofer lines in the solar spectrum and the general characteristics of metallic spectra as generated in the arc. For the complete analysis of complicated spectra a larger instrument with a wave-length scale is necessary.

Tube Spectroscope. The grating, lens, and slit may be arranged equally well in telescoping metal tubes of similar lengths to those employed in the construction of the box spectroscope, if one can do the metal work and prefers a metal instrument.

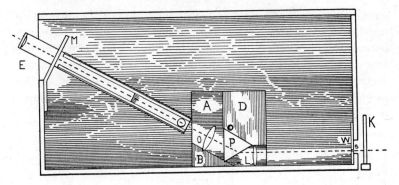

FIGURE 21
Prism Spectroscope

PRISM SPECTROSCOPE. The 45° prisms used in tanks and periscopes do not have sufficient dispersion for spectroscopic use without high telescopic magnification, and probably few of them have sufficient finish to give good results under such high power. Even a single 60° prism is of little use without telescopic magnification. A precisely ground, highly polished, bubble-free, flint glass prism, which will usually cost about $10, affords the nucleus for a very good instrument.

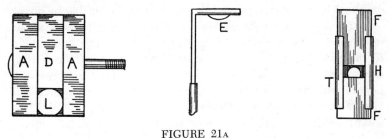

FIGURE 21A
Parts for the Prism Spectroscope

The mounting described here is novel but effective, and is adapted to home construction. The prism, collimating lens, and telescope objective are all held between two boards, each 5 inches square. Grooves are sawed in both boards, Fig. 21A, for holding opposite edges of the lenses. The prism P, is placed between the lenses, as illustrated, Fig. 21, and held by a notched triangular board, B, Fig. 21B, of the same thickness as the prism, and the rectangular board D, one corner of which is sawed off at the proper angle to hold the prism P in proper position, as shown in Fig. 20.

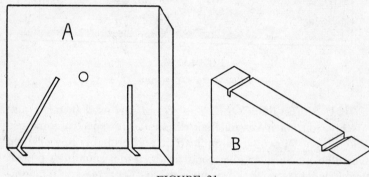

FIGURE 21B
Other Parts for the Prism Spectroscope

Since prisms will not all have the same refractive index, the exact angle is best determined by experiment, but it will be about as illustrated. Set up the lenses and prism and shift them about until the light rays enter and leave the prism at the same angle to the prism faces and pass through the centers of the lenses; then mount them in this fixed relative position between the boards with the lens edges in the slits and the corner pieces B and D tacked in place. After all parts are secure, bolt the two square holding boards A together and stand the unit on edge.

Case. The lens-prism unit is then placed in a box about 19 inches long, 6 inches wide, and 9 inches high. The top is removable, and the rear wall has an opening about 2 inches square near the top. The ocular-lens E is mounted at the free end of an arm whose opposite end is bolted to the side wall of the case near the prism block. The length of the arm should about equal the sum of the focal lengths of the eye- and objective-lenses. For length adjustment the arm may consist of a flat rod or strip sliding in a sheath. The strip is bent at right angles near the end and the ocular lens mounted upon it. The scale M is mounted upon a strip of metal attached to the rear wall and projecting out between the ocular and the telescope objective O, in such a manner as to be viewed simultaneously with the spectrum and directly alongside of it.

The prism block is moved back from the slit S until the collimator L is about its focal length from it, but is not bolted to the side of the case until repeated trials have shown exactly where the spectral lines are sharpest and the background blackest.

Slit. A window about an inch in diameter is drilled in the front of the case and the slit fitted over it. The track T for the slit may consist of a piece of tin with opposite edges folded over to hold the blades and with a hole H punched in the center. Two strips of brass F with sharpened and carefully honed edges are slid into the track from opposite ends so as to leave a narrow cleancut slit across the hole H. The track is attached in front of the window W with the slit arranged horizontally. A sheet of glass K is placed in front of the slit for protection.

An enlarged image of the arc is projected upon the slit by a condenser lens, and the light rays then pass in succession through the slit, the collimator, the prism, the telescope objective and the eye-lens. The prism changes the direction of the beam as shown by the dotted line. The focal length of the ocular-lens E is about one inch, the objective lens O, 10-inches, and the collimator L, six inches. The prism is equilateral with 60° angles. One looks down into the spectroscope in much the same manner as into a microscope. Since the slit is horizontal the carbons are horizontal also. Very good results are possible with simple non-achromatic lenses if they are carefully centered and focused.

The Scale. Using a magnifying glass, mark off a row of dots uniformly spaced along the edge of a piece of paper and number every fourth one. Clamp this slip to the scale holder, which is bent so as to be seen clearly through the eye-lens as it sweeps over the spectrum. Next, burn a fragment of lepidolite in the arc or moisten the tips of the carbons in lithium chloride and write down the numbers of the dots that coincide with the spectral lines red, orange and blue. Again using the magnifying glass, make a new scale with dots spaced like the lithium lines. Give them their approximate Ångstrom numbers: * red 6700, orange 6100 and blue 4600. Similarly mark the place for the sodium doublet at 5900, and enough other lines to complete the scale. If the scale is painstakingly made and carefully adjusted, the spectroscope is equipped to do real analytical work.

Masks. A piece of black cardboard with a hole about ½ inch in diameter centered over the objective lens will greatly improve the definition of the spectral lines by eliminating the spherical aberration originating at the edges of the lens. Even with achromatic lenses it is customary to block off the light coming from too near the edges.

Grating Modification. A highly efficient grating spectroscope may be made merely by substituting a 15,000-line grating for the prism in the above mounting.

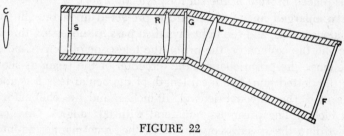

FIGURE 22
Plane Grating Spectrograph

PLANE GRATING SPECTROGRAPH. With care, very fine spectrograms may be made with plane gratings. Fig. 22 shows the arrangement of condenser *C*, slit *S*, collimator *R*, grating *G*,

---

* The Ångström values of lines, as given here, are not quite exact, but are within the limits of accuracy of the apparatus used.

focusing lens L and film F. Either direct or alternating current may be used. The vertical arrangement of carbons, slit and grating lines is usually preferred. A cup is drilled in the tip of the lower carbon to hold the sample. Achromatic lenses give better spectrograms.

An enlarged image of the arc is focused upon the slit S by the condenser C in such a manner that the images of the luminous carbons are beyond the ends of slit, so as not to fog the plate. The collimator R is at its focal length from the slit S and for sharp focus over the full length of the film F the film will have to be held on a curve. This curve may be determined by marking

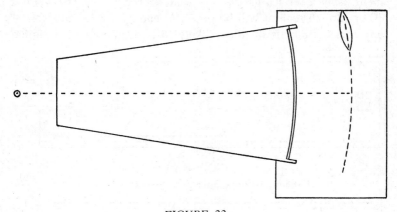

FIGURE 22A
The Film Curve

the path of an ocular lens as it goes from color to color all the time keeping the spectrum in focus, by geometrically, or experimentally, determining the center of the circle traveled by the ocular, subtracting the focal length of the ocular from the radius of the circle, and finally drawing a new arc concentric with the first with the shortened radius. If the film is arranged on this arc the spectral lines will be in focus throughout its entire length. The path of light is turned at an angle by the grating, and this angle will be greater with a 25,000-line than with a 15,000-line grating.

THE CONCAVE SPECTROMETER.* For use in rapid visual analysis the writer prefers the concave-grating spectrometer. Especially if one is planning to construct his own instrument he should give consideration to this type because of its great simplicity, high efficiency, low cost, ease of operation, high dispersion and the long accurate scale.

The necessary optical parts are the slit, the concave grating and the ocular. Parts to be built are the arc, the case, mounts for the optical parts and the scale. The total cost for such an instrument using a replica grating need not exceed $15 or $20. With a 10 or 12 inch scale it will be powerful enough for the analysis of any rocks, minerals or alloys likely to be encountered. Original

FIGURE 23
Concave Grating Spectrometer (Top View)

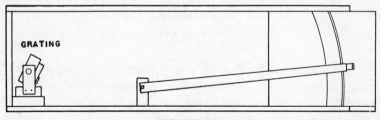

FIGURE 23A
Concave Grating Spectrometer (Side View)

* The term spectroscope has been used in this book to cover an instrument without a scale, spectrometer an instrument with a scale for measuring the spectrum, spectrograph an instrument for photographing the spectrum.

gratings are expensive, but good replicas give splendid results for either visual or photographic analysis.

Of several possible mountings the writer has found the following most suitable for rapid results. A replica of one meter focus (radius of curvature) and 15,000 lines per inch will be taken for illustration, but any other concave replica may be mounted in a similar manner. The focal length of the grating may be tested by placing a white cardboard beside an unfrosted lamp bulb but so as to be shielded from its direct light, and by reflecting the image of the filament upon the cardboard by means of the concave grating. If the distance between grating and cardboard, and between grating and lamp, is approximately 39 inches, the focus of the grating will be one meter. In this mounting the distance between arc and slit is 4 inches.

Parts are all to be mounted upon a baseboard an inch thick, 10 inches wide and 4 feet long. The grating holder consists of a little base three inches square and a pivoted block of the same width held between two uprights from the former. A shelf for holding the grating projects from the bottom of the pivoted block; the front edge of the shelf is turned up to form a clamp for securely holding it. The grating holder is attached to one end of the baseboard of the spectrometer by a single screw to permit horizontal rotation. In this way the spectrum reflected from the grating may be directed upon the ocular adjacent to the scale. The block which holds the grating is also secured to each upright by a single screw to permit the grating to be tipped slightly forward in such a manner that the spectral lines will correspond with the proper lines of the scale.

Window. A partition is placed between grating and arc 35 inches from the grating and 4 inches from the arc. This contains a window half an inch in diameter on an exact level with both grating and arc. The carbons are mounted horizontally in holders rising from a pedestal, and the pedestal is attached to the baseboard of the spectrometer by a single screw to permit moving the arc back and forth to keep it in alignment with window and grating.

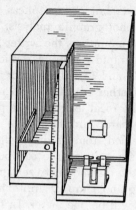

FIGURE 23B
Concave Grating Spectrometer (Perspective View)

Slit. The slit may consist of two new safety razor blades edge to edge, Fig. 23c, held in position by a strip of metal, about the width of one blade, folded over them at each end. A quarter-

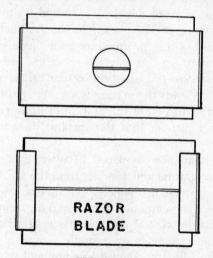

FIGURE 23c
Spectroscopic Slit Made from Razor Blades (Front and Back Views)

inch hole is punched in the middle of the strip for the passage of light between the blades. The blades project beyond the strip both above and below and are held firmly, but not too tightly by the end folds to permit adjustment until the slit width which gives best results is found. The slit is now placed over the window so as to exclude all but the flattened beam of light entering between the sharp edges of the blades. To permit adjustments of the grating and focusing of the arc upon it, the slit mechanism should be easily removable so that the full light of the arc may pass through the window. A shielding partition is also placed between the arc and the ocular. The scale holder as illustrated is attached to this partition, Fig. 23B.

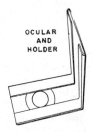

FIGURE 23D
Ocular (Lens and Holder)

Ocular. An eyepiece consisting of a single lens with a focal length of about one inch, is attached at right angles to one end of an arm half a meter long. The other end of the arm is bolted either to the adjacent side of the case or to a standard rising from the base as illustrated. The bolt should be provided with a wing nut to permit the regulation of tension. The rotation of the arm permits it to sweep over the entire visible spectrum as one looks through the ocular. The arm may be of adjustable length, consisting near the end of a flat rod and sheath, or of telescoping tubes. The distance between grating and ocular will be about 40 inches.

Scale. Mark the positions of the ocular as it is pushed up and down, and drawn in and out, which bring out clearly the different parts of the spectrum. This will determine the curve of the scale, for on the partition separating it from the arc these markings on the wall will fall upon the arc of a circle; if the ocular arm is bolted at the center of curvature of this arc, it may thereafter be of fixed length with the scale always in focus.

The scale support will take the form of an arc with a radius one inch shorter than that traversed by the ocular. It may be sawed from a board an inch thick and attached to the partition. A paper scale may be used, but to prevent expansion from moisture it should be glued to a metal strip. It should also be held somewhat loosely at top, bottom and back to permit longitudinal adjustment for articulation with the lines of the spectrum, since the instrument will constantly expand and contract enough to throw the scale out of fine adjustment.

The case is about a foot high and closed except for the half end holding the ocular; the interior is painted flat black. The top may be removable from the case, or the whole case removable from the baseboard, as preferred. The arc occupies one corner of the baseboard and is shut off from the interior compartment by partitions, except for the slit and for a series of quarter-inch holes bored along the curve of the scale for illumination if it is not otherwise sufficiently lighted.

Operation. The light from the arc passes through the slit to the concave grating; there it is dispersed as a magnified spectrum, the red spectral lines going to the top of the scale, the violet to the bottom. The spectrum is then magnified about ten times further by the ocular and may be viewed in any desired portion. The arc is placed so, conveniently at the hand of the operator, that it may be manipulated even while the sample is under observation. One soon learns to adjust carbons, raise or lower the rock in the arc or shift the arc from side to side without once taking his eyes from the spectrum. The compact arrangement greatly speeds the task of analysis, all of which may be completed without leaving one's seat. Just to the rear of the arc is placed a blue cobalt glass through which may be observed the conditions of the arc while starting or operating it.

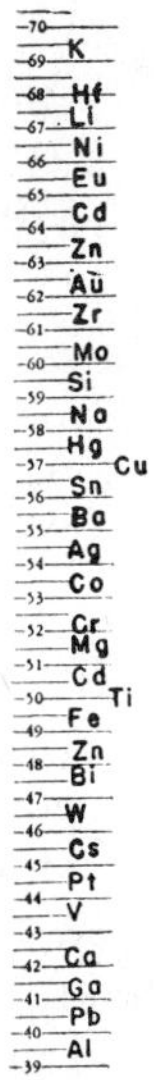

FIGURE 23E
Calibrated Scale

Calibration. The scale is made in much the same manner as for the prism instrument. First a rough scale with arbitrary numbers but fine markings is prepared by use of a magnifying lens. This is clamped upon the holder and written notes taken upon the positions of such spectral lines as K-6911; Li-6707, 6104 and 4603; Pb-6002, 5201 and 5005; and Fe-4202. The wave-lengths of all these lines fall near even hundreds, and all are strong and easily found with little chance of confusion. Since the spacings with this mounting are nearly the same for all colors, remaining numerals are easily supplied. A new scale is now made by transferring the marks from the arbitrary scale to the new one and writing in the proper numerals to denote Ångstrom units.

Checks. The scale should now be inserted in the holder and

52   *Manual of Spectroscopy*

carefully aligned with the sodium doublet at 5896. Then it should be checked throughout its entire length by such lines as: K-6939, Ca-6717, Ba-6595, Ba-6496, Ti-6303, Ba-5800, Cu-5700, Ca-5602, Fe-5404–5, Fe-5302, Cr-5204, Cu-5105, Ti-4999, Fe-4924, Zn-4810, Cu-4704, Sr-4607, Fe-4404, Fe-4307, Fe-4203, Ca-4226, Pb-4057, Al-3944, Al-3961. When lines the whole length of the spectrum check closely, the scale can be depended upon thereafter for accuracy.

Scale Symbols. The scale may also contain the symbols of the elements in the regions of the spectrum where their most sensitive or characteristic lines are to be found. Cu(350) may be placed a little further from the edge of the scale than the number 57 (for 5700). This will indicate that one of the best lines for the identification of copper will appear at that point, if copper is present in the sample, and that it will be a strong line of intensity 350. The scale is large enough to indicate in this manner one or two of the best lines for the identification of each of the elements, facilitating the performance of rapid and thorough work. A magnifying glass should be used for writing in the symbols, and they should be in red ink to avoid possible confusion with scale numbers.

Element List. The following lines may be marked. Do not write in the Ångstrom numbers; they are given only to show where to put the symbols on the scale; write in only the symbols followed by their respective intensities at the indicated spots on the scale. Cs(500) 6973, K(500) 6939, K(300) 6911, Sm(1300) 6861, Gd(500) 6846, Hf(100) 6819, Cs(500) 6723, Li(3000) 6707, Yb(1000) 6667, Ba(200) 6527, Ta(500) 6485, Tm(400) 6460, Co(1000) 6450, Zn(1000) 6362, Rb(1000) 6298, Au(700) 6278, Ni(600) 6256, Lu(500) 6221, Rb(800) 6206, Zr(300) 6143, Ba(2000) 6141, Zr(300) 6134, Zr(500) 6127, Ca(100) 6122, Li(2000) 6103, F(Band 200)* 6064, Mo(300) 6030, Mn(100) 6021, Mn(100) 6016, Mn(100) 6013, Pb(40) 6001, Yt(Band 300) 5987 [M.I.T. tables give Yt(300) 5987.6], Eu(1000) 5966, Si(50) 5948, Ho(200) 5948–33, U(125) 5915, Na(5000) 5896–90, Hg(600)

---
* For explanation of bands see Chapter IV.

Spectroscopes

5790–69, V(200) 5706, Cu(350) 5700, Co(600) 5647, Sn(50) 5631, Ca(15) 5601, Bi(500) 5552, Ba(1000) 5535, Mo(200) 5506, U(60) 5492, Ag(1000) 5465, Fe(300) 5455–46 Tl(5000) 5350, Cb(400) 5344, Co(800) 5343–2–1, Re(500) 5275, Cu(700) 5218, Cr(500) 5204–6–8, Mg(500) 5183–72–67, C(Band) 5165, Cu(600) 5153–05, Cd(1000) 5085, Th(40) 5028, Ti(200) 4999, Fe(1000) 4925, Re(2000) 4889, W(50) 4843, Mn(400) 4825, Zn(400) 4810, Au(200) 4792, Mn(400) 4754–62–66, Bi(1000) 4722, Zn(400) 4722, W(200) 4659, Sr(1000) 4607, Li(800) 4602, Cs(1000) 4593, Be(15) 4572, Ru(1000) 4554, Sn(500) 4524, In(5000) 4511, Pt(100) 4498, Pr(125) 4477, Fe (1000) 4404, Ni(1000) 4401, V(200) 4379, Rh(1000) 4374, Hg(3000) 4358, La(800) 4333, Ce(50) 4320, Cr(5000) 4254–74, Ca(500) 4226, Ge(200) 4226, P(300) 4222, Pd(500) 4212, Pr(100) 4189, Ga(2000) 4172, La(600) 4077, Pd(500) 4087, Pb(2000) 4057, Er(35) 4008, Ce(60) 4012, Tb(100) 4033, Sc(100) 4023, Dy(400) 4000, Al(3000) 3961–44.

Mask. The rulings of the grating, like the slit and the carbons, must be mounted horizontally so as to throw a vertical spectrum with the red lines at the top and the violet at the bottom. If the lines are fuzzy or distorted, check the slit and grating to see that they are in proper position. If there is no improvement, try masking the grating, that is, placing in front of it a piece of black paper or cardboard from which a half-inch square has been cut out. By using only a portion of the grating surface, results may be greatly improved. Try different parts of the grating and determine what part gives the best lines. Though part of the light is lost, it is better to use only that part which gives clear lines, for resolution is more important than brilliance. A spectroscope of this type should separate lines less than two angstroms apart.

Condenser Lens. If for any reason, the mounting previously described does not give satisfactory results, a condensing lens may be placed between arc and slit. First remove the slit and adjust the lens to throw an image of the arc upon the face of the grating. While the window is open, tilt the grating to send the first order spectrum to the ocular and turn it so that the spectrum will be adjacent to the scale. The lens may be placed either inside or outside the case with similar results, provided that the image of the arc is projected upon the grating. The slit is then replaced and the spectrum examined. The spectral lines will be lengthened and the black strip of field widened by use of the lens.

Orders of Spectra. A prism yields but a single spectrum, but a grating produces several. If one holds a plane grating close to the eye and looks toward an unfrosted light bulb, or any similar source of light, there will appear two equally bright continuous spectra on opposite sides of the light. These are the primary spectra. Fainter and longer secondary spectra will also be seen on both sides of the light but farther from it. If the grating is of good quality, spectra of the third, the fourth, and even possibly of higher orders may be seen still farther away on both sides of the light, but they will be faint and so long that their colors overlap.

Concave gratings show the same phenomena. By tilting the grating farther forward the secondary spectrum will be thrown upon the ocular instead of the primary. Although this is sometimes done to increase dispersion in making spectrograms, the loss in brilliance is usually too great to be worth while in visual work, especially since a new scale would also be required.

Concave Replicas. Replicas of original concave gratings are made by pouring a solution of collodion in amyl acetate upon the original grating and allowing it to dry for several days. It is then loosened by immersing in water and the collodion film mounted upon a concave mirror of the same radius of curvature. Taking off replicas eventually destroys the rulings of the original, but twenty or thirty good gratings may be obtained. The replicas carry such exact reproductions of the original rulings that their spectra are often scarcely distinguishable from those of the master grating. The collodion may be cleaned with a dry brush, but alcohol would instantly destroy it, and even water might loosen the replica from the mirror.

Refinements. Baffles as well as black paint may be used to prevent reflections inside the case, especially for photographing the spectra. Diaphragms may be used at the slit to block off the bright continuous spectra of the carbons. Adjusting screws for turning the grating, for tilting it, for regulating the position of the scale, for controlling the carbons and the like may be added at the pleasure of the builder, but the essentials of a working spectrometer have been given.

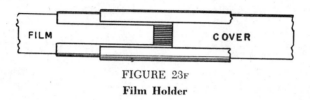

FIGURE 23F
Film Holder

SPECTROGRAPH. For photographing the spectrum, a strip of film such as is used for motion pictures is placed directly alongside the scale. The film holder may consist of two sheaths made of sheet metal, one inside the other, with a flat slide covering the opening of the inner sheath. The film is held in the inner sheath, and the slide removed for the exposure, which is usually from five to ten seconds. A thick black cloth is thrown over the open end of the case to exclude all light but that entering

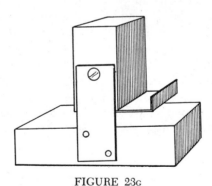

FIGURE 23G
Grating Holder

the slit. Long exposure will fog the bright lines and bring out the weak ones. Short exposure will make sharp bright lines but very faint weak ones. The spectrogram may be made longer than the scale if a quartz condensing lens, or no lens, is used, for the spectrum may be photographed well into the ultraviolet if desired.

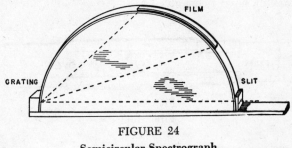

FIGURE 24
Semicircular Spectrograph

SEMICIRCULAR SPECTROGRAPH. A very simple form of spectrograph may be made by sawing from plyboard two semicircles each having a diameter equal to the focal length of the concave grating, which (unlike a concave mirror) is equal to its radius of curvature. Mount the two boards upright upon a baseboard side by side and a few inches apart. Cover the dome with a long strip of sheet metal. Allow the base to project at one end to accommodate the arc and provide a spectroscopic slit in the curved strip about two inches above the base, as illustrated.

Mount the grating in a little compartment at the far end of the base against a window in the curved strip. Provide also a long narrow window near the top of the spectrograph so that a strip of film may be placed over the opening, and provide a cover over the film. The grating is tilted back to throw the beam of light (shown as dotted lines in the Figure) received from the slit upward to the film.

A sliding ocular consisting of a single lens of one inch focus, may be moved along an inch above the window for inspecting the spectrum, and removed when the film is to be used. In this compact mounting the slit, grating and film are all in one vertical plane on the circumference of a circle, an arrangement producing uniform dispersion throughout the spectrum and making computation of wave lengths very simple. A mask at the slit adapted for lateral movement will enable the operator to make comparison spectrograms side by side. This mounting is not so convenient for visual analysis as the one previously described, but it is very effective for photographing the spectrum. When the spectogram is examined with a magnifying lens the spacings will appear to be about the same as those of the spectrum when examined through the ocular.

Chapter IV

SPECTROSCOPIC ANALYSIS

1. Qualitative Analysis

Qualitative analysis determines the elements of which a sample is composed. Quantitative analysis determines their relative proportions. An understanding of the several types of phenomena to be seen in the spectroscope will facilitate spectroscopic analysis.

Continuous Spectra are bands of brilliant colors which shade into each other without interruption, and which are produced by incandescent solids such as electric-light filaments or the white-hot tips of the arc carbons. These spectra sometimes appear alongside the useful spectra, but the more completely they are masked out the better.

Absorption Spectra are those resulting from the selective absorption of light by gases, liquids or solids. The gases surrounding the sun absorb the light of the vaporized elements in it and, as a result, a dark line spectrum is produced. Examination of the more than 20,000 fine black lines show the presence in the sun of at least 60 of the elements found upon earth. Many elements burned in the laboratory arc produce similar clouds of gas which cause the *reversal* from brilliantly colored to black lines.

Window glass, while transmitting visible light, absorbs or blanks out most of the ultraviolet radiation. Colored transparent media blot out certain fields of color, sometimes in characteristic patterns. Weak potassium permanganate solution, for example, in a flat flask placed between the spectroscope and a brilliant source of light, will give five narrow shadow bands in the blue region; uranium nitrate crystals will give a group in the green; cobalt glass will give distinct shadow bands in the red, orange and green.

Some colorless hydrocarbons may be identified by dark bands produced in the ultraviolet region upon a photographic plate, or by the extent and degree of their transparency to ultraviolet radiation. For such liquids, however, infrared and mass spectrographs are more efficient as previously explained.

Band Spectra. Bands are flutings or groupings of very fine lines which crowd closer and closer together at one end until they culminate in a more or less brilliant head. They are believed to be the result of molecular vibrations and are illustrated by the numerous cyanogen and carbon monoxide bands to be seen in the empty carbon arc. These bands are as definitely placed in the spectrum as the spectral lines and therefore are sometimes useful as spotters. Generally, however, they only clutter up the spectral field, and fog spectrograms. Fortunately they will often wane, as the bright-line spectra wax, and at times disappear entirely. Calcium fluoride produces several very brilliant bands which afford the only visible arc means for the detection of fluorine. The bands of yttrium and some other elements may also be used for indentification purposes.

Dust Lines. Still other lines often appear in the spectroscope picture and, to the confusion of the novice, are sometimes even more pronounced than those of the spectrum. These are black lines caused by imperfections in the optical parts, by dust particles, or by the rough edges of the slit. However, since they are always at right angles to the spectral lines they can be distinguished from them easily.

Bright-line Spectra. The lines most used in spectral analysis are the brightly colored ones appearing upon the dark spectroscopic field. They are produced by the substances being vaporized in the electric arc, spark or flame, or in the excited vapors of the Geissler tube. Some of these lines are brilliant, and some faint. Sometimes they are so numerous as to crowd the spectroscopic field, sometimes they are few and scattered. Each element has its own assortment of colored lines by which it may be recognized.

The angstrom is the unit of length used in measuring light

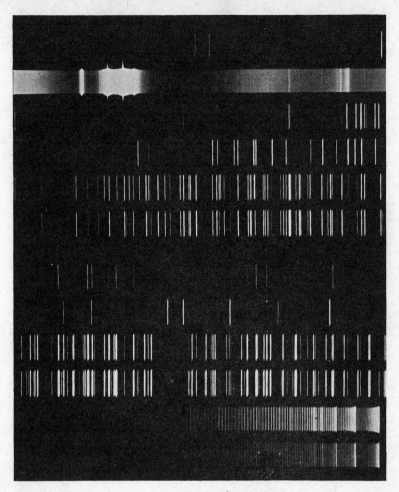

FIGURE 25

Types of Spectra

The first five spectrograms above are of the spark covering the ultraviolet region from 2880 to 3070 Å. The one at the top shows Pb-3068, the next Mg-2928, 2936. The third shows an aluminum group, Al-3050 to Al-3066 and other fainter aluminum lines. The fourth shows the nickel lines of the region, and the fifth iron. The sixth line shows the iron lines of the arc for comparison.

The seventh spectrum shows the aluminum lines from 3700 to 3900 Å (spark). All following lines are of the same region. The eighth spectrogram is of nickel (spark) the ninth of iron (spark); the tenth of iron (arc); and the last two show the cyanogen bands of the carbon arc.

Spectroscopic Analysis 63

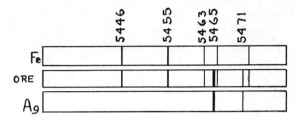

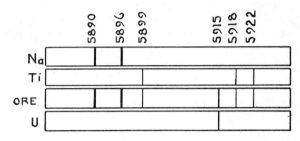

FIGURE 27
Comparison Diagrams

The first of the accompanying diagrams shows a group useful for recognizing silver. At the top is shown the iron spectrum; at the bottom the silver spectrum; in the middle is the spectrum of the ore containing both metals.

The second figure shows the most easily identified uranium line, with sodium and titanium lines at the top, the uranium line at the bottom, and the ore with all three elements in the middle. Such diagrams as these are easily made and are very helpful.

Comparison Spectra. If the instrument contains no scale one must burn sodium, lithium, copper or some other known element in the arc along with the unknown so that its lines may serve as a scale. The relative distances between the known and the unknown lines must be carefully judged and then the wave-length calculated. Reference to the table-chart at the end of the book will then disclose the identity of the unknown. Suppose, for instance, that when an unknown and copper are burned simultaneously in the arc a very bright line turns the copper triplet into a quadruplet by adding a line toward the violet. Suppose further that the distance that separates it from the copper group is equal to about half the distance between the two copper lines 5100-50, i.e., 25 angstroms. Reference to the table-chart will show a bright cadmium line at that place. Before making a positive identification, however, look up that metal and check one or two additional lines.

Use of Scale. If the instrument has considerable dispersion and is provided with a scale, the task of identification is greatly simplified, for the scale then takes the place of the comparison lines and shows the wave-length without computation. One has but to read the scale number indicated by the spectral line and refer directly to the table-chart. Line brilliance must always be taken into consideration, for often faint lines of one element and brilliant ones of another are to be found at almost the same place in the spectrum.

Much more rapid and accurate visual analysis can, of course, be made with grating than with prism instruments, because the scale spaces on the former are all equal and easily estimated, while with the latter they grow smaller and smaller toward the red end of the spectrum. The dispersion of the grating is usually greater than that of the prism, and it is possible to calibrate a long scale much more accurately than a short one.

Persistent Lines. If a sample containing any element is sufficiently diluted by the addition of other substances, some of the lines of that element will disappear. If the dilution is carried far enough all lines will disappear, but it has been found that certain lines always persist longer than the others; therefore, the few that remain longest are called *persistent*. These lines are frequently, but not always, the most intense lines of the spectrum. Tests have shown that the persistent lines of cobalt, lead, titanium, strontium, chromium, aluminum, and copper will be found in the spectrum even when the concentration of these elements is one one-thousandth of one per cent of the weight of the sample;

those of lithium, sodium, calcium and strontium will be found at still lower concentrations. In tables they are usually indicated by the letter P.

Intensities. Under the same conditions of excitation, spectral lines of the same element and of different elements differ widely in brilliance. There are two common systems in use for indicating intensities. In one the numeral 10 is used to indicate maximum intensity or brilliance, in the other 10,000. In both systems decreasing numerals indicate decreasing intensity with the numeral 1 indicating the weakest lines. The 10,000 to 1 system has been used in this book, since the relative intensities, which are very important in spectroscopic analysis, are much more reliably indicated by it.

Determination of a Single Element. To determine the presence or absence of a single element in a sample it is, of course, unnecessary to hunt for all its lines. The chapter on Characteristic Lines will indicate which lines are best for this purpose, and as soon as two or three of these are found, or found wanting, it is unnecessary to look further.

Complete Determination. Determination of the complete composition of a rock or sample is more difficult, but it does not involve the consideration of every line that appears in the spectrum. One soon becomes so familiar with the usual line groupings that he can run down the spectrum very rapidly and thoroughly, but the beginner will have to undertake a systematic examination in order not to overlook anything. The accompanying table of key lines has been prepared to simplify this task. In it are given the strategic lines of all elements appearing in the arc. At least two lines of each element are given, one or both of which are sensitive or persistent ones; the other may be merely a prominent line in a convenient spot for identification. If these key lines are absent, it is unnecessary to look further, for if sensitive lines are missing the element is not there. When the key lines of an element seem to be present, turn to the chapter on Characteristic Lines and check one or two other lines found there. While a single line is often sufficient, and sometimes in the case of mere traces, the only one present, yet it is desirable to find other lines of the element if possible. A lone line must be carefully checked with known neighbors to make it trustworthy. Line intensities are indicated by the numbers in parentheses.

TABLE 5

Key Lines

Wave-length, Å	Element	Intensity	Wave-length, Å	Element	Intensity
6939	K	(500)	5915	U	(125)
6923	Ru	(300)	5900	Cb	(200)
6911	K	(300)	5895	Na	(5000)
6861	Sm	(800)			
			5889	Na	(9000)
6857	Gd	(200)	5769	Hg	(600)
6846	Gd	(500)	5700	Cu	(350)
6818	Hf	(100)	5700	Sc	(400)
6723	Cs	(500)	5632	Sn	(50)
6707	Li	(3000)			
			5556	Yb	(1500)
6438	Cd	(2000)	5535	Ba	(1000)
6435	Yt	(150)	5522	Ce	(100)
6362	Zn	(1000)	5476	Ni	(400)
6298	Rb	(1000)	5465	Ag	(1000)
6278	Au	(700)			
			5414	Er	(50)
6249	La	(300)	5353	Co	(500)
6146	Ti	(400)	5352	Co	(500)
6143	Zr	(300)	5350	Tl	(5000)
6134	Zr	(300)	5291	F	(200)
6127	Zr	(500)			
			5218	Cu	(700)
6103	Li	(2000)	5209	Ag	(1500)
6079	Sb	(20)	5206	Cr	(500)
6030	Mo	(300)	5183	Mg	(500)
6004	Lu	(400)	5172	Mg	(200)
6001	Pb	(40)			Band
5948	Si	(50)	5167	Mg	(100)
5921	Ho	(200)	5153	Cu	(600)

Spectroscopic Analysis

TABLE 5 (continued)
Key Lines

Wavelength, Å	Element	Intensity	Wavelength, Å	Element	Intensity
5085	Cd	(1000)	4384	V	(125)
5017	Th	(50)	4374	Rh	(1000)
5007	Ti	(200)			
			4358	Hg	(3000)
4999	Ti	(200)	4303	Nd	(100)
4921	La	(500)	4302	W	(60)
4920	La	(500)	4278	Tb	(200)
4889	Re	(2000)	4254	Cr	(5000)
4832	Sr	(200)			
			4241	U	(40)
4823	Mn	(400)	4226	Ca	(500)
4810	Zn	(400)	4226	Ge	(200)
4794	Os	(300)	4225	Pr	(50)
4772	Zr	(100)	4222	P	(300)
4722	Bi	(1000)			
			4212	Pd	(500)
4681	Ta	(200)	4211	Dy	(200)
4680	W	(150)	4205	Eu	(200)
4680	Zn	(300)	4201	Rb	(2000)
4669	Ta	(300)	4186	Ce	(80)
4659	W	(200)			
			4179	Pr	(200)
4607	Sr	(1000)	4172	Ga	(2000)
4572	Be	(15)	4101	In	(2000)
4555	Cs	(2000)	4058	Cb	(1000)
4554	Ba	(1000)	4057	Pb	(2000)
4524	Sn	(500)			
			4044	K	(800)
4518	Lu	(300)	4033	Ga	(1000)
4511	In	(5000)	4030	Mn	(500)
4454	Ca	(200)	4000	Dy	(400)
4442	Pt	(800)	3988	Yb	(1000)
4424	Sm	(300)			
			3961	Al	(3000)
4420	Os	(400)	3951	Nd	(40)
4404	Fe	(1000)	3944	Al	(2000)
4399	Ir	(400)	3911	Sc	(150)

TABLE 5 (continued)

Key Lines

Wavelength, Å	Element	Intensity	Wavelength, Å	Element	Intensity
3906	Er	(25)	3185	V	(500)
3902	Mo	(1000)	3183	V	(200)
			3134	Hf	(80)
3891	Ho	(200)	3067	Bi	(3000)
3874	Tb	(200)			
3762	Tm	(200)	3064	Pt	(2000)
3719	Fe	(1000)	3039	Ge	(1000)
3646	Gd	(200)	2881	Si	(500)
			2852	Mg	(300)
3601	Th	(8)	2534	P	(100)
3600	Yt	(100)			
3524	Ni	(1000)			
3515	Ni	(1000)	2497	B	(500)
3498	Ru	(500)	2496	B	(300)
			2478	C	(400)
			2427	Au	(400)
3460	Re	(1000)	2385	Te	(600)
3453	Co	(3000)			
3434	Rh	(1000)	2383	Te	(500)
3404	Pd	(2000)	2349	As	(250)
3311	Ta	(300)	2348	Be	(2000)
			2288	As	(250)
3220	Ir	(100)	2068	Sb	(300)

Characteristics. Hydrogen and lithium are among the elements having the fewest spectral lines. Cerium and uranium have the most numerous lines and the weakest. Europium has the greatest number of unusually strong lines, although sodium, potassium and rubidium each have a few lines which are still stronger. The iron spectrum, having over 5000 lines distributed quite evenly over the entire range, is often used as a comparison spectrum for spotting the lines of the other elements.

With small spectroscopes one considers large line groups or even the appearance of the spectrum as a whole, while with larger instruments one examines small groups, or even single lines.

The following table has been arranged for testing the dispersion of instruments and for practice in judging the spacing of lines in angstrom units.

TABLE 6

DOUBLETS

Element	Wave-Length, Å	Intensity	Spacing Å	Element	Wave-Length, Å	Intensity	Spacing Å
Ti	5036.4	125 ⎫	0.5	Fe	5434.5	300 ⎫	5
Ti	5035.9	125 ⎭		Fe	5429.6	500 ⎭	
Fe	6137.6	100 ⎫	1	Na	5895.9	5000 ⎫	6
Fe	6136.6	100 ⎭		Na	5889.9	9000 ⎭	
Ti	5038.4	100 ⎫	2	Fe	4619.2	100 ⎫	8
Ti	5036.4	100 ⎭		Fe	4611.2	200 ⎭	
Fe	5269.5	800 ⎫	3	Mg	5183.6	500 ⎫	11
Fe	5266.5	500 ⎭		Mg	5172.6	200 ⎭	
Ti	5020.0	100 ⎫	4	Fe	5191.4	400 ⎫	20
Ti	5016.1	100 ⎭		Fe	5171.5	300 ⎭	

The following lines and line groups are of interest for their intensity or arrangement (Intensities are given in parentheses.):

SINGLETS—BRILLIANT

Lithium	6708 (3000)	
Cadmium	6438 (2000)	
Rubidium	6299 (1000)	
Bismuth	4722 (1000)	
Thallium	5350 (5000)	
Cesium	4593 (1000)	and 4555 (2000)
Strontium	4607 (1000)	
Barium	6141 (2000)	and 5535 (1000)
Silver	5465 (1000)	and 5209 (1500)
Indium	4511 (5000)	
Lead	4057 (2000)	
Europium	5966 (1000)	
Nickel	4410 (1000)	

SINGLETS—WEAK
 Silicon   5948 (50)
 Gold   6278 (700)
 Molybdenum   6030 (300)
 Tin   5631 (50) and 4525 (500)
 Beryllium   4572 (15)
 Calcium   4226 (500)

DOUBLETS
 Potassium   6911-39 (300-500)
 Sodium   5890-96 (9000-5000)
 Iron   5446-55 (300-300) and 4925 (1000)
 Mercury   5770-90 (600)
 Cerium   4523-27-28 (35-50-30)
 Lanthanum   4921-22 (500-500)
 Aluminum   3944-61 (2000-3000)
 Tantalum   4661-69 (300-300)

TRIPLETS—FIRST 2 LINES ONE HALF AS FAR APART AS LAST TWO
 Calcium   6103-22-62 (80-100-40)
 Manganese   6113-16-22 (100-80-80)
 Chromium   5204-6-8 (400-500-500)
 Magnesium   5167-72-83 (100-200-500)
 Copper   5105-53-5218 (500-600-700)
 Molybdenum   5507-33-70 (200-200-200)
 Zinc   4680-4722-4810 (400-400-300)

QUADRUPLETS—IRREGULAR SPACING
 Cobalt   5339-41-42-43 (100-300-800-600)
 Manganese   4754-62-83-4823 (400-100-400-400)
 Chromium   4226-54-75-89 (500-5000-4000-3000)

QUINTUPLETS—REGULAR SPACING
 Titanium   4981-91-99-5000-07 (300-200-200-200-200) also 4525
 Vanadium   4379-85-90-95-4400 (200-125-80-60-60)

SEPTUPLETS
 Calcium   5582-88-90-94-98-5601-02 (20-35-15-35-35-15-15)

Outer Electrons. The spectra of the alkali elements, lithium, sodium, potassium, rubidium and cesium are similar. All of them consist of few but very strong widely scattered lines and their tendency to doublets is striking. Their atomic structures are such that they have a single electron in their outer shells. The spectra of the elements with two electrons in their outer shells, i.e., magnesium, calcium, strontium, and barium have similarities which are almost as striking. All are easily excited in the 110-volt arc and have distinctive groupings and lines well scattered through all the colors, but with little crowding. Metals with few electrons in the outer shell, such as silver, gold, copper, zinc, cadmium and mercury also have similar spectra of few lines.

The elements with outer shells about half filled (See Figure 1.), such as barium, aluminum, gallium, indium, carbon, silicon, tin, lead, phosphorus, arsenic, antimony, bismuth, zirconium, etc., have spectra containing few lines, most of them weak. The elements oxygen, fluorine, chlorine, bromine and iodine all require the Geissler tube for their excitation. The metallic elements vanadium, colodium, tantalum, chromium, molybdenum, tungsten, manganese and rhenium all show similar complicated spectra with a great wealth of both strong and weak lines. Iron, nickel, cobalt, and the platinum group show similarly complex spectra. The rare earths have spectra made up of many weak lines.

The number of electrons in the outer shell of the atom has the greatest influence upon its spectrum. The more electrons there are in the shell, the more complex is the spectrum. The inert gases with complete shells have spectra averaging about 1200 lines each, except for helium which has only 110 lines with an intensity greater than (1).

2. Quantitative Analysis

With the spectroscope, it is possible to make visual quantitative estimates of considerable accuracy and photographic analyses of very great accuracy, especially when the quantities to be considered are small.

The spectrum of every element consists of lines of unequal intensity, some strong, some weak, and some very weak. A large number of characteristic lines, both bright and faint, indicates the presence of a high percentage of the element in the sample. The brighter the lines and the greater their number, the higher the concentration of the element. For example, the spectrum of celestite, containing 48 per cent of strontium, will show virtually all the strontium lines, both strong and weak. However, if only a trace of strontium is present, as is the case with the average rock, the only line of the strontium spectrum to be seen will be the very persistent blue line Sr-4607.

Intensity Factors. In making quantitative estimates one must take into consideration both the method of excitation and the quality of the spectroscope. An increase in amperage will tend to increase brilliance; prisms sometimes give brighter lines than gratings; increased dispersion always results in decreased intensity; voltage changes almost always affect intensities. Quantitative tests must consequently be made under the same standard conditions for trustworthy results.

Steady Visible Lines. The stronger spectral lines of copper, iron, magnesium, zinc and many other metals tend to become steady and dependable when the concentration of these metals is about 1 per cent; below that concentration they tend to fade and fluctuate. In contrast, the persistent lines of titanium and vanadium become bright and steady when the concentration of these elements is only a small fraction of 1 per cent. A striking example of the unpredictable nature of a spectrum is the case of argon, which was found to require a concentration of 37 per cent in a certain discharge tube and at a certain pressure to bring out any of its lines. Arc lines are much more dependable.

Complete Ignition. It should be noted that the metals in the arc are melted and vaporized in succession according to their boiling points. The spectra of some metals, therefore, appear and disappear more quickly than others. For quantitative results most spectrographers take a small sample, such perhaps as 50 mg, and burn it completely. The visual analyst, however, must depend upon average brilliance over a longer period of time, with frequent replenishment of the sample.

Reversed Lines. One of the simplest ways to judge quantities is to note the reversal of spectral lines. In low concentration all the visible lines of an element will be colored; but in higher proportion many elements throw off clouds of gas which absorb the light produced in the arc and turn certain lines black upon a luminous field. The percentage at which such reversals occur may be observed and used in quantitative estimates. The alkaline elements having only one or two electrons in the outer shell all have very strong reversed lines. Other elements such as molybdenum, iron, gold, copper and bismuth have no reversed visible lines, although most have ultraviolet ones.

Weak Lines. Since the weak lines of an element appear only with increasing concentration, they provide a means for estimating higher concentrations. As an example, the only visible calcium line at 0.01 per cent concentration in a carbon arc operating on 110 volts at 800 watts will be Ca-4226. At 1 per cent the group Ca-4425-4434-4455 appears, at about 2 per cent Ca-6102-6122-6169 and then Ca-5588-5590-5594-5601-5602. At 8 per cent all the calcium lines have become strong but Ca-4226 is not yet reversed. At 20 per cent this line is strongly reversed and the others are at full intensity.

Graduated Standards. For visual quantitative estimates a graduated series of standard powders for each element to be judged is made up. These powders are burned one at a time and careful notes taken; finally the notes are tabulated. By the aid of such tables one can soon learn to differentiate between ore and worthless rock and to estimate the relative proportions of the elements present in the sample. Chemicals of sufficient purity for this purpose are easily obtainable from chemical dealers or from drug stores.

TABLE 7

SYMBOLS AND PROPERTIES OF THE ELEMENTS *

Element	Symbol	Atomic Number	Atomic Weight	Number of Outer Electrons	Outer Shell Number	Melting Point	Boiling Point	Valence
Actinium	Ac	89	227	3	7	. .	. .	. .
Alabamine	Ab	85	221	31	6	. .	. .	1, 3, 5, 7
Aluminum	Al	13	26.97	3	3	660	1800	3
Antimony	Sb	51	121.76	15	5	630	1380	3, 5

TABLE 7 (continued)

Element	Symbol	Atomic Number	Atomic Weight	Number of Outer Electrons	Outer Shell Number	Melting Point	Boiling Point	Valence
Argon	A	18	39.94	8	3	—189.2	—185.7	0
Arsenic	As	33	74.91	15	4	Sub	615	3, 5
Barium	Ba	56	137.36	2	6	850	1140	2
Beryllium	Be	4	9.02	2	2	1350	1530	2
Bismuth	Bi	83	209.00	29	6	271	1450	3, 5
Boron	B	5	10.82	3	2	2000	2550	3
Bromine	Br	35	79.92	17	4	—7.2	58.8	1
Cadmium	Cd	48	112.41	12	5	320.9	767	2
Calcium	Ca	20	40.08	2	4	810	1170	2
Carbon	C	6	12.01	4	2	Sub	3500	4
Cerium	Ce	58	140.13	4	6	640	1400	3, 4
Cesium	Cs	55	132.91	1	6	26.4	670	1
Chlorine	Cl	17	35.46	7	3	6101.6	—34.6	1
Chromium	Cr	24	52.01	6	4	1615	2200	2, 3, 6
Cobalt	Co	27	58.94	9	4	1480	2900	2, 3
Columbium	Cb	41	92.91	5	5	1950	3300	3, 5
Copper	Cu	29	63.57	11	4	1083	2300	1, 2
Dysprosium	Dy	66	162.46	12	6	. .	. .	3
Erbium	Er	68	167.2	14	6	. .	. .	3
Europium	Eu	63	152.0	9	6	. .	. .	3
Fluorine	F	9	19.00	7	2	—223	—187	1
Gadolinium	Gd	64	156.9	10	6	. .	. .	3
Gallium	Ga	31	69.72	13	4	29.75	2000	3
Germanium	Ge	32	72.60	14	4	958	2700	4
Gold	Au	79	197.2	25	6	1063	2600	1, 3
Hafnium	Hf	72	178.6	18	6	1700	3200	3
Helium	He	2	4.00	2	1	—271.9	—268.9	0
Holmium	Ho	67	163.5	13	6	. .	. .	3
Hydrogen	H	1	1.008	1	1	—259	—258.8	1
Illinium	Il	61	146	7	6	. .	. .	3
Indium	In	49	114.76	13	5	155	1450	3
Iodine	I	53	126.92	17	5	113.5	184.4	1
Iridium	Ir	77	193.1	23	6	2350	4800?	3, 4
Iron	Fe	26	53.84	8	4	1535	3000	2, 3
Krypton	Kr	36	83.7	18	4	—169	—151.8	0
Lanthanum	La	57	138.92	3	6	826	1800	3
Lead	Pb	82	207.21	28	6	327.5	1620	2, 4
Lithium	Li	3	6.94	1	2	186	1200	1
Lutecium	Lu	71	175	17	6	. .	. .	3
Magnesium	Mg	12	24.32	2	3	651	1110	2
Manganese	Mn	25	54.93	7	4	1260	1900	2, 4, 6, 7
Masurium	Ma	43	. .	7	6	2300	. .	. .
Mercury	Hg	80	200.61	26	6	—38.8	365.9	1, 2
Molybdenum	Mo	42	95.95	6	5	2620	3700	3, 4, 6
Neodymium	Nd	60	144.27	6	6	840	. .	3
Neon	Ne	10	20.18	8	2	—248.6	—245.9	0
Neptunium	Np	93	. .	7	7	. .	. .	. .
Nickel	Ni	28	58.69	10	4	1452	2900	2, 3
Nitrogen	N	7	14.01	5	2	—209.9	—195.8	3, 5
Osmium	Os	76	190.2	22	6	2700	5300?	2, 3, 4, 8

TABLE 7 (continued)

Element	Symbol	Atomic Number	Atomic Weight	Number of Outer Electrons	Outer Shell Number	Melting Point	Boiling Point	Valence
Oxygen	O	8	16.00	6	2	—218.4	—183	2
Palladium	Pd	46	106.7	10	5	1555	2200	2, 4
Phosphorus	P	15	31.02	5	3	44.1	280	3, 5
Platinum	Pt	78	195.23	24	6	1755	4300	2, 4
Plutonium	Pu	94	. .	8	7	. .	. .	. .
Polonium	Po	84	. .	30	6	. .	. .	. .
Potassium	K	19	39.09	1	3	62.3	760	1
Praseodymium	Pr	59	140.92	5	6	940	. .	3
Protoactinium	Pa	91	231	5	7	. .	. .	. .
Radium	Ra	88	226.05	2	7	960	1140	2
Radon	Rn	86	222	32	6	—71	—61.8	0
Rhenium	Re	75	186.31	21	6	3000	. .	. .
Rhodium	Rh	45	102.91	9	5	1955	2500?	3
Rubidium	Rb	37	85.48	1	5	38.5	700	1
Ruthenium	Ru	44	101.7	8	5	2450	2700?	2, 4, 6, 8
Samarium	Sm	62	150.43	8	6	1300	. .	3
Scandium	Sc	21	45.10	3	4	1200	2400	3
Selenium	Se	34	78.96	16	4	220	688	2, 4, 6
Silicon	Si	14	28.06	4	3	1420	2600	4
Silver	Ag	47	107.88	11	5	960.5	1950	1
Sodium	Na	11	22.99	1	3	97.5	880	1
Strontium	Sr	38	87.63	2	5	800	1150	2
Sulfur	S	16	32.06	6	3	119	4446	2, 6
Tantalum	Ta	73	180.88	19	6	2850	4100	5
Tellurium	Te	52	127.61	16	5	452	1390	4, 6
Terbium	Tb	65	159.2	11	6			3
Thallium	Tl	81	204.39	27	6	303.5	1650	3
Thorium	Th	90	232.12	4	7	1845	3000	4
Thulium	Tm	69	169.4	15	6			3
Tin	Sn	50	118.70	14	5	231.9	2260	2, 4
Titanium	Ti	22	47.90	4	3	1800	3000	3, 4
Tungsten	W	74	183.92	20	6	3400	4727	6
Uranium	U	92	238.07	6	7	1150		4, 6
Vanadium	V	23	50.95	5	4	1710	3000	3, 5
Virginium	Vi	87	224?	1	7			1
Xenon	Xe	54	131.3	18	5	—112	—107.1	0
Ytterbium	Yb	70	173.04	16	6	1800		3
Yttrium	Yt	39	88.92	3	5	1490	2500	3
Zinc	Zn	30	65.38	12	4	419.4	907	2
Zirconium	Zr	40	91.22	4	5	1700	2900	4

* Recent investigation made the existence of masurium, illinium, alabamine and virginium doubtful. By means of the cyclotron new radioactive elements have been found to occupy their places in the atomic series. These are as follows: tachnetium, Tc, 43; cyclonium, Cy, 61; astatine, At, 85; and francium, Fr, 87. In addition, the following new transuranium elements have been discovered: americium, Am, 95 and curium, Cm, 96.

TABLE 8

Content of Certain Elements in Their Most Common Compounds

Compound	Formula	Element	Content %
Aluminum Phosphate	$AlPO_4$	Aluminum	22
Arsenic Oxide	As_2O_3	Arsenic	76
Barium Sulfate	$BaSO_4$	Barium	59
Beryllium Oxide	BeO	Beryllium	36
Bismuth Trioxide	Bi_2O_3	Bismuth	90
Boric Acid	H_3BO_3	Boron	17
Cadmium Sulfate	$CdSO_4$	Cadmium	54
Calcium Oxide	CaO	Calcium	71
Ceric Oxide	CeO_2	Cerium	81
Chromium Sulfide	Cr_2S_3	Chromium	25
Cobalt Oxide	CoO	Cobalt	79
Cuprous Sulfide	Cu_2S	Copper	80
Ferric Oxide	Fe_2O_3	Iron	70
Lead Chromate	$PbCrO_4$	Lead	64
Magnesium Carbonate	$MgCo_3$	Magnesium	29
Manganese Dioxide	MnO_2	Manganese	63
Molybdenum Oxide	MoO_2	Molybdenum	77
Nickel Monoxide	NiO	Nickel	78
Potassium Chlorate	$KClO_3$	Potassium	32
Silicon Dioxide	SiO_2	Silicon	47
Silver Chloride	$AgCl$	Silver	57
Sodium Chloride	$NaCl$	Sodium	38
Strontium Chloride	$SrCl_2$	Strontium	58
Barium Sulfate	$BaSO_4$	Sulfur	14
Stannic Chloride	$SnCl_4$	Tin	46
Titanium Dioxide	TiO_2	Titanium	60
Uranium Dioxide	UO_2	Uranium	88
Vanadium Dioxide	V_2O_2	Vanadium	77
Zinc Oxide	ZnO	Zinc	80
Zirconium Dioxide	ZrO_2	Zirconium	74

Procedure. The standard powder is mixed in varying known proportions with a spectroscopically inert substance so that the brilliance of the lines at the different concentrations may be compared. The best all around adulterant for this purpose is silica, since it has few visible lines and these very faint. Commer-

Spectroscopic Analysis 77

cial powdered silica will usually have too much lime and other impurities to be satisfactory, but if one selects and pulverizes his own pure white quartz fragments, the spectral field can be kept virtually free from all lines but those of the standard. It is unnecessary to use quartz crystals; milky quartz or common white opal will serve as well. Aluminum oxide also makes a fairly good diluent.

The procedure may be illustrated by the test for zinc. Since pure metallic zinc is difficult to divide into particles fine enough to be thoroughly mixed with diluent, zinc oxide will serve better. From the Table it will be seen that zinc oxide contains approximately 80 per cent zinc, and so will serve as the 80 per cent standard.

Weigh out 250 mg of the oxide and add enough pulverized, sifted quartz to make 2 g. The original zinc oxide was 80 per cent zinc; the same amount of zinc is now contained in 8 times the weight of powder, making a 10 per cent concentration. Grind and mix the powder thoroughly and place it in an envelope labeled 10 per cent zinc. Next, weigh out 250 mg of the 10 per cent mixture and add enough quartz to make up 2.5 g, giving a 1 per cent mixture. Similarly, prepare 0.1 per cent and 0.01 per cent mixtures. A very small quantity of filings from a sheet of metallic zinc will give a 100 per cent standard.

First, burn the sample containing the lowest percentage of zinc, then the next stronger and so on up to the 100 per cent sample. Take careful notes upon each spectrum. The findings will depend upon the type of arc, type of instrument, voltage, etc., but will run something as follows:

Concentration of Sample:

0.01 per cent, Lines at 4680, 4722 and 4810 Å come and go.
0.1 per cent, Same lines; still feeble, sometimes failing.
1.0 per cent, Blue lines faint, but steady; Zn-6362 strong, in flashes.
5.0 per cent, All the persistent lines strong and steady.
10.0 per cent, Zn-6362 now surpasses the blue lines in intensity.
80.0 per cent, Little increase in intensity as compared with the 10 per cent sample.
100.0 per cent, Same as with the 80 per cent.

When chemicals or ores contain more than about 10 per cent of the element to be estimated, they must, as a rule, be diluted until some of the spectral lines disappear or appreciably weaken. To estimate, for example, the percentage of calcium in calcite, equal amounts of calcite and quartz are weighed and pulverized together in a mortar. The test shows lines much too near full intensity for an estimate. Again dilute the mixture, or any part of it, by the addition of an equal weight of silica. If the lines of the diluted powder are still too strong, the dilution continues until the lines are weakened enough to indicate a concentration of about 5 per cent. The original percentage may then be computed.

Multiple Standards. Powders consisting of equal concentrations of several different elements will save much time in assembling quantitative data, and will emphasize the different sensitivities by simultaneous presentation. Of course, elements with very complicated spectra should be taken singly or perhaps in pairs to avoid confusion of lines. If the elements chosen are those commonly found in rocks, or those coming up frequently for estimation, one will very soon be able to make practical quantitative estimates.

A standard powder for the comparison of barium, magnesium, manganese, potassium and calcium may be prepared as follows: the barium sulfate molecule ($BaSO_4$) consists of one barium atom, one sulfur atom and four oxygen atoms. The molecular weight is the sum of the atomic weights, Ba(137.36) plus S(32.06) plus four times O(16) gives the total molecular weight 233.42. The ratio of the weight of the whole molecule to that of the barium atom is therefore 233.42 $\div$ 137.36, or 1.69. This means that in 1.69 g of $BaSO_4$ there will be exactly 1 g of barium.

The magnesium carbonate molecule ($MgCO_3$) consists of one atom each of magnesium (24.32) and carbon (12), and three atoms of oxygen (16); its molecular weight is 84.32. The ratio of the molecular weight to that of the magnesium atom is therefore 84.32 $\div$ 24.32 or 3.46. In the same way the ratio of the molecular weight of manganese dioxide (MnO_2) to the weight

Spectroscopic Analysis 79

of the manganese atom is 1.58; in the case of potassium chloride (Kcl) and potassium, the ratio is 1.90, and in that of calcium oxide (CaO) and calcium, 1.39. Now weigh out the following amounts:

BaSO₄	1.69g
MgCO₃	3.56g
MnO₂	1.58g
KCl	1.90g
CaO	1.39g
Total	10.2g

This mixture may be used directly as a 10 per cent standard for each metal in it, since it contains 1 g each of barium, magnesium, manganese, potassium and calcium in 10 g of the whole mixture; from it 5 per cent, 2 per cent, and 1 per cent standards may be made by adding silica in the manner previously described.

Thus, elements may be observed singly, in pairs, or in any desired grouping. If notes are taken with each observation and tabulated, quantitative work may soon be done with precision and dispatch.

In the following table is given the ratio of the molecular weight of the compound to the weight of the element to be determined, that is: $\dfrac{\text{Molecular weight}}{\text{Atomic weight}}$.

TABLE 9

RATIO OF THE MOLECULAR WEIGHT OF A COMPOUND TO THE ATOMIC WEIGHT OF ONE OF ITS CONSTITUENT ELEMENTS

Compound	Formula	Molecular Weight	Element	Atomic Weight	Ratio
Aluminum oxide (corundum)	Al_2O_3	101.94	Aluminum	26.97	1.89
Antimony sulfide (stibnite)	Sb_2S_3	339.70	Antimony	121.76	1.39
Arsenous oxide (arsenolite)	As_2O_3	197.86	Arsenic	74.93	1.32
Barium monosulfide	BaS	169.42	Barium	137.36	1.23
Barium sulfate (barite)	$BaSO_4$	233.42	Barium	137.36	1.69
Beryllium-Aluminum silicate (beryl)	$Be_3O_{18}Al_2Si_6$	537.36	Beryllium	9.02	19.85
Bismuth dioxide	BiO_2	241.00	Bismuth	209.00	1.15
Boron oxide (boric anhydride)	B_2O_3	69.64	Boron	10.82	3.14
Cadmium sulfide (greenockite)	CdS	144.47	Cadmium	112.41	1.28

TABLE 9 (continued)

Compound	Formula	Molecular Weight	Element	Atomic Weight	Ratio
Calcium oxide (lime)	CaO	56.08	Calcium	40.08	1.39
Calcium chloride (hydrophilite)	$CaCl_2$	110.99	Calcium	40.08	2.76
Calcium carbonate (calcite)	$CaCO_3$	100.08	Calcium	40.08	2.49
Ceric oxide	CeO_2	172.13	Cerium	140.13	1.22
Cesium chloride	CsCl	168.27	Cesium	132.81	1.26
Chromic oxide	Cr_2O_3	152.02	Chromium	52.01	1.46
Cobaltic chloride	$CoCl_3$	165.31	Cobalt	58.94	2.80
Cupric sulfide (covellite)	CuS	95.63	Copper	63.57	1.5
Calcium fluoride (fluorite)	CaF_2	78.08	Fluorine	19.00	4.1
Indium monoxide	InI	241.68	Indium	114.80	2.1
Ferrous oxide	FeO	71.84	Iron	55.84	1.28
Ferrous carbonate (siderite)	$FeCO_3$	115.84	Iron	55.84	2.07
Ferric oxide (hematite)	Fe_2O_3	159.68	Iron	55.84	1.43
Lead monoxide (litharge)	PbO	223.22	Lead	207.22	1.07
Lead sesquioxide	Pb_2O_3	426.44	Lead	207.22	1.11
Lithium chloride	LiCl	42.40	Lithium	6.94	6.1
Magnesium sulfate	$MgSO_4$	120.38	Magnesium	24.32	4.95
Magnesium carbonate (magnesite)	$MgCO_3$	84.32	Magnesium	24.32	3.46
Magnesium oxide (periclase)	MgO	40.32	Magnesium	24.32	1.65
Manganese silicate (rhodonite)	$MnSiO_3$	130.99	Manganese	54.93	2.38
Manganese dioxide (pyrolusite)	MnO_2	96.93	Manganese	54.93	1.58
Mercuric sulfide (cinnabar)	HgS	232.67	Mercury	200.61	1.1
Molybdenum disulfide (molybdenite)	MoS_2	160.12	Molybdenum	96.0	1.66
Nickel monosulfide (millerite)	NiS	90.75	Nickel	58.69	1.54
Phosphorus arsenide	PAs	105.95	Phosphorus	31.02	3.38
Platinum dioxide	PtO_2	227.23	Platinum	195.23	1.16
Potassium chloride (sylvite)	KCl	74.56	Potassium	39.10	1.90
Potassium chlorate	$KClO_3$	122.56	Potassium	39.10	3.13
Potassium nitrate (saltpeter)	KNO_3	110.11	Potassium	39.10	2.58
Rubidium hydroxide	RbOH	102.45	Rubidium	85.44	1.19
Selenium monobromide	Se_2Br_2	318.23	Selenium	79.2	2.01
Silicon dioxide (quartz)	SiO_2	60.06	Silicon	28.06	2.14
Silver chloride	AgCl	143.34	Silver	107.88	1.32
Sodium chloride (halite)	NaCl	58.45	Sodium	22.99	2.54
Strontium nitrate	$Sr(NO_3)_2$	211.65	Strontium	87.63	2.41
Thallium cyanide	TlCN	230.40	Thallium	204.39	1.12
Thorium sulfide	ThS_2	296.24	Thorium	232.12	1.27
Stannic oxide (cassiterite)	SnO_2	150.70	Tin	118.70	1.26
Titanium dioxide (rutile)	TiO_2	79.90	Titanium	47.90	1.66
Tungsten dioxide	WO_2	216.00	Tungsten	184.0	1.17
Uranium trioxide	UO_3	286.14	Uranium	238.14	1.20
Vanadium dioxide	V_2O_2	133.90	Vanadium	50.90	2.62
Zinc oxide	ZnO	81.38	Zinc	65.38	1.24
Zirconium silicate (zircon)	$ZrSiO_4$	183.28	Zirconium	91.22	2.01

TABLE 10 *

Quantitative Intensities

	Line		0.01%	0.1%	1%	2%	4%	8%	20%
Barium	6141	(1000)	T	D	C	B	A	A	
	5519	(200)		D	C	C	B	A	
	4554	(1000)		D	C	B	B	A	
	4934	(400)		D	C		B	A	

Spectroscopic Analysis

TABLE 10 (continued)

	Line		0.01%	0.1%	1%	2%	4%	8%	20%
Bismuth	4722	(1000)				D	C	B	
Calcium	5598	(35)		D	C	C	B	B	
	4454	(200)			C	C	B	A	
	4434	(150)			D	C	B	A	
	4426	(500)	D	C	C	B	A	A	R
Chromium	5208	(100)	D	C	B	B	A	A	
	5206	(500)	D	C	B	B	A	A	
	5204	(500)	D	C	B	B	A	A	
	4289	(800)	D	C	B	B	A	A	
Copper	5700				T	D	C	C	B
	5218	(700)		T	D	C	C	B	B
	5153	(600)			T	D	C	B	B
Fluorine	5291	Band					B	B	
Iron	5455	(300)				D	D	C	C
	5446	(300)				D	D	C	C
	4903	(500)		D	D	C	C	B	B
	4383	(1000)		C	C	C	C	B	B
	4415	(600)		C	C	C	C	B	B
	4404	(1000)		C	C	C	C	B	B
Lead	6002	(40)					D	C	B
	4057	(2000)					D	C	B
Magnesium	5183	(500)		T	D	C	B	B	B
	5172	(200)				C	B	B	B
	4073	(8)						T	D
Manganese	6021	(80)				D	C	C	B
	6016	(80)				D	C	C	B
	6013	(100)				D	C	C	B
	4823	(400)		T	D	C	C	B	A
	4783	(400)		T	D	C	C	B	A
Potassium	6939	(500)					C	B	
	6911	(300)					C	B	

TABLE 10 (continued)

	Line		0.01%	0.1%	1%	2%	4%	8%	20%
Potassium	4047	(200)					C	B	
	4044	(800)					C	B	
Sodium	5889	(9000)	D	C	C	B	A	R	
	5895	(5000)	D	C	C	B	A	R	
	5688	(300)				T	D	C	
	5682	(80)					D	C	
Tin	5631	(50)			T	D	C	C	C
	4525	(500)			T	D	C	C	C
Titanium	5007	(200)	D	C	B	B	A	A	
	4999	(200)	D	C	B	B	A	A	
	4991	(200)	D	C	B	B	A	A	
	4759	(100)			D	D	C	C	
	4758	(125)			D	D	C	C	
Tungsten	4843	(50)			D	C	C	C	
	4660	(200)			D	C	B	B	
Zinc	6362	(1000)		T	T	D	C	B	
	4810	(400)	T	D	C	C	B	B	
	4722	(400)	T	D	C	C	B	A	
	4680	(300)	T	D	C	C	B	B	
Zirconium	6143	(300)		D	D	C	B	B	
	6134	(300)		D	D	C	B	B	
	6127	(500)		D	D	C	B	B	

° Explanation of Table.

Table 10 shows the intensities (numerals in parentheses) of some of the lines most often encountered in the analysis of rocks, at the percentages indicated. These intensities will vary with different instruments, but the table will suggest the manner in which visual estimates may be made. In securing the above values a concave grating with 15000 lines per inch and focal length of one meter was used. The meanings of the symbols used are:

    T—trace, mere occasional flashes of the line
    D—Very weak line, sometimes disappearing entirely
    C—Steady dependable line, only moderately strong
    B—Strong dependable line of considerable brilliance
    A—Line of maximum intensity for the element
    R—Reversed line, appearing black against a colored field

The Spectrogram. Photographing the spectrum makes possible highly accurate quantitative analyses, based on the principle that high concentrations make brighter spectral lines and brighter lines make denser spectrograms. The amount of material taken for the sample, the time of exposure and the current supply are all carefully regulated. The sample is placed upon the lower of two vertical electrodes and is usually burned completely. Direct-current arc, alternating-current arc, and spark are all in common use. Several different methods are possible, some comparing the spectrogram of the sample with a master spectrogram, others measuring the density of the lines recorded on the films. (See spectrograms at the beginning of the chapter.)

Paired Lines. Probably the simplest generally applicable method is that of comparing lines of equal intensity. First a series of standard concentrations of a single element is prepared (as described under visual analysis) and the spectra of these standards photographed, then the lines of all other elements are compared with these standards, on the assumption that lines of the same rated intensity strengthen and weaken in unison as the percentage in the sample changes. For the standards, mixtures of increasing concentrations, up to 10 per cent are made of some such element as manganese and their spectra photographed side by side, as many on one film as possible, each being labeled with the percentage of manganese used.

A spectrogram of the sample to be analyzed is then made under the same controlled conditions. Should the sample contain copper among the elements to be determined, any recognized line of copper, for instance Cu-4704 is selected. Its rated intensity (200) is then determined from the tables. Next, a manganese line of equal intensity is selected from the table of manganese lines on page 84.

The manganese line Mn-5341 has the required intensity and so is selected for comparison. All that remains is to search through the standard spectrograms of manganese for the Mn-5341 line that matches in density the Cu-4704 line of the sample. The percentage of manganese required to produce the standard line is known; therefore the concentration of copper is determined, for it is the same.

The percentage of zinc, should there be any in the sample,

TABLE 11
INTENSITY OF MANGANESE LINES

Wavelength, Å	Intensity	Wavelength, Å	Intensity	Wavelength, Å	Intensity	Wavelength, Å	Intensity
4649	(2)	4363	(20)	4408	(60)	5341	(200)
4712	(5)	4365	(25)	4766	(80)	4034	(250)
4755	(10)	4586	(30)	4343	(100)	2576	(300)
4594	(12)	4645	(40)	4502	(125)	4754	(400)
4730	(15)	4305	(50)	4709	(150)	4030	(500)

may be determined by matching such lines of equally rated intensity as Zn-4911 (15) and Mn-4730 (15); the percentage of vanadium by matching V-5401 with Mn-4343, since both are rated in the tables with intensity (100). The percentage of any other element in the sample, or in any other sample, may be found in the same manner by matching one of its lines with the appropriate manganese line.

The dependability of these findings is limited by the accuracy of the rated intensities, which have been taken from the *Wavelength Tables* prepared with government aid at the Massachusetts Institute of Technology. Although based upon carefully controlled spectrograms, they do not pretend to absolute accuracy. The closer together in the spectrum the lines are chosen, the more dependable will be the results, since differences in film sensitivity in the red, violet, and ultraviolet regions make widely scattered lines less reliable. In producing the lines, a standard direct-current electric arc on a 220-volt line was used, with current stabilizers.

A further limitation on using this method is the fact that two lines of equal intensity do not always fade exactly in unison. For utmost precision each pair of lines used would have to be checked at several concentrations. A persistent line of one element should not be matched with a non-persistent line of another, since a weak persistent line will sometimes outlast a stronger non-persistent one as the percentage is reduced. Lines of medium intensity usually give best results. Accuracy may be checked by comparing with a second standard, e.g., copper. In spite of limitations the method is so simple that it will be found practicable where too precise results are not required.

Precision Method. Although this effect is regarded as negligible by many spectroscopists, it has been found that one element will occasionally affect the line intensities of another in a spectrogram. Manganese, silicon and zinc have been found to suppress aluminum lines, and calcium to enhance the lines of potassium and magnesium. Errors due to this and most other causes may be avoided by making a preliminary spectrogram of the sample and from it estimating the proportions of each element found in it. A standard is then synthesized containing the same elements in the estimated proportions. If the lines match, then the percentage of each element is determined; if they do not all match, then new standards are synthesized until they do. The use of the same elements in both standard and sample is very common in industrial analysis, where the exact percentage of each metal is so important in certain alloys. Since the same metals appear in each operation, a series of standards containing those elements in varying known proportions is prepared and the alloys compared with the standards. As a check against possible variations in the arc, a standard is often photographed alongside the sample. When lines are matched for equal density the two spectrograms are usually placed upon a glass plate illuminated from beneath. The lines are then examined through a magnifying glass which also shows a wave-length scale. At other times densitometers, such as have been described in Chapter III are used in comparing line densities.

Density Readings. Percentages are sometimes computed directly from spectrograms made of the sample to which has been added a bit of some element of known characteristics. Since the density of an exposed and developed film, however, is not di-

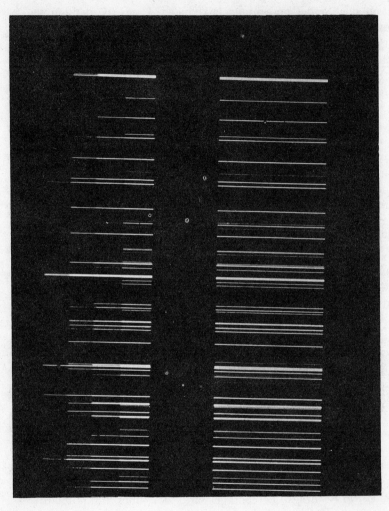

FIGURE 28

Spectrograms with and without a logarithmic sector. The spectrum

rectly proportional to the intensity of the light used in the exposure, nor the percentage of the element to the blackness of the line, it is necessary to use logarithmic functions in the computation. Calculating boards have been designed for this purpose which do a large part of the work mechanically, using graphs.

Logarithmic Sectors are frequently rotated between the arc and the slit to shut out part of the light. These are disks with rims partially cut away to mask part of the light from one end of all the spectral lines so that they taper to points and the weaker lines become shorter than the strong ones. The percentage of the element is then a logarithmic function of the length of the spectral line.

Comparative Accuracy. In general, spectroscopic analyses for concentrations of less than 1 per cent are more accurate than chemical analyses. Between 1 per cent and 10 per cent the two methods are about equally reliable, with the speed factor usually on the side of spectroscopy. In concentrations over 10 per cent, although the spectroscope is sometimes applicable, chemical quantitative methods are usually preferred. The limits of spectroscopic precision, however, have not yet been reached.

Chapter V

THE SPECTROSCOPE IN MINERALOGY

Since minerals are classified primarily according to chemical composition, the spectroscope is invaluable in the examination of specimens. Crystal form, hardness, microscopic structure, magnetic properties, refractive index and other characteristics are all important, and no one instrument or method is self-sufficient. Each must be used to supplement the others.

Tests Used in Mineral Identification

Flame Colors. The blowpipe, so popular with miners and mineralogists of the past, disclosed that many substances burn or vaporize with distinctive colors. Salt yellows the entire flame, strontium imparts an orange flare, copper tints it green, zinc blue, and lithium red. When the prospector saw these colors, he knew which elements caused them. The spectroscope makes the same use of colors, but can detect the complex colors of the iron, nickel, or cobalt flame as readily as simple ones and thus reveals completely which metals are to be found in rocks and minerals.

Rapid Testing. Spectroscopic tests are so rapid that each crystal, vein, or colored spot of a specimen can be tested for composition within a few minutes. They show what elements are present and in what approximate proportions. If a complete quantitative analysis is required, the spectroscope will serve to direct the chemical procedure. At each step it will show the completeness of the separation, and in the end the purity of the final products.

Spectroscopic Sampling. In ore analysis much importance is placed upon selecting truly representative samples. The amount of gangue that must be removed in following the vein must be taken into consideration in estimating the value of the ore per ton. The samples are crushed after collection and thoroughly mixed. The crushed material is then divided into quarters and ¼ of each quarter taken for further pulverization and mixing. This quartering, mixing and grinding continues until the final size of the sample is suitable for analysis. In this way, a small sample becomes truly representative of the whole vein. The same method should be used in quantitative spectroscopic analysis. Where only qualitative results are sought, it will often be more satisfactory to burn different bits or corners of the rock directly without crushing.

Rock Composition. It is very helpful to know the average composition of the rock crust of the earth, since this will indicate what elements one may ordinarily expect to find in common rocks, and in what proportions. This average, computed by F. W. Clarke * (Data of Geochemistry) from hundreds of samples collected from all lands, is as follows:

TABLE 12

AVERAGE COMPOSITION OF THE LITHOSPHERE

Element	Percentage	Element	Percentage	Element	Percentage
Oxygen	46.5	Hydrogen	0.14	Fluorine	0.03
Silicon	27.6	Phosphorus	0.12	Strontium	0.02
Aluminum	8.1	Manganese	0.09	Vanadium	0.02
Iron	5.1	Carbon	0.09	Nickel	0.02
Calcium	3.6	Sulfur	0.06	Lithium	0.01
Sodium	2.8	Chlorine	0.05	Copper	0.01
Potassium	2.6	Chromium	0.05	Zinc	0.004
Magnesium	2.1	Barium	0.04	Lead	0.002
Titanum	0.6	Zirconium	0.04	All others	0.3

The following Table gives the composition, or average composition of some of the typical rocks found in the United States:

* See bibliography.

TABLE 13
Composition, or Average Composition, of Typical American Rocks

Rock	SiO_2 %	Al_2O_3 %	FeO %	CaO %	Na_2O %	K_2O %	MgO %	H_2O %	TiO_2 %	MnO %	BaO %	SrO %	Li_2O %	Cr_2O_3 %	NiO %	CO_3 %	Zn %	SO_3 %	P_2O_5 %
Granite	74.3	14.4	1.1	0.6	1.8	6.6	0.1	1.0	0.1	Tr*	Tr		Tr						Tr
Rhyolite	73.2	12.7	1.1	0.6	1.9	5.2	0.2	5.0	0.1	Tr	Tr		Tr						
Obsidian	75.2	14.1	1.8	0.8	3.9	3.6	0.1	0.4										0.1	Tr
Dacite	68.1	15.3	3.2	3.0	4.2	3.1	0.1	2.7	0.2	Tr	0.1	Tr							
Andesite	57.0	18.5	6.0	4.3	5.0	3.7	2.3	2.3	0.3	0.2									0.4
Diorite	57.7	16.3	5.6	5.5	5.0	0.8	5.5	2.7	0.5	0.1	0.1	Tr	Tr						0.2
Basalt	47.9	18.2	10.8	9.9	2.8	0.3	8.2	1.1	0.6	Tr			Tr						0.2
Gabbro	55.9	13.5	8.6	8.7	2.4	1.7	6.5	1.7	0.6	0.1	Tr		Tr						0.3
Clay	71.1	12.5	6.0	0.9	2.2	1.6	0.4	4.4	0.5	Tr						0.4			Tr
Sandstone	77.7	4.4	1.4	5.5	0.5	1.3	1.2	1.6	0.3	Tr	0.1	Tr	Tr			5.0	0.1	0.8	0.1
Chert	98.9	0.7	0.3	0.1			Tr												
Shale	55.4	14.8	5.7	6.0	1.8	2.7	2.7	5.5	0.5	Tr	0.1		Tr			4.6			0.2
Limestone	5.3	0.8	0.5	43.0	0.1	0.3	7.9	0.8	0.1	0.1			Tr			41.0		0.1	
Serpentine	41.0	0.5	6.0	0.4	0.3	0.2	37.0	12.0		0.1		Tr		0.4	0.1				Tr
Mica Schist	70.0	15.2	3.2	1.6	3.2	3.5	1.5	1.0	0.5	0.1	0.1		Tr						0.1

* Tr = Traces.

Application of the Spectroscope in Mineral Classification

In some mineral classes, the spectroscope is indispensable, since by no other means can the metallic minerals be so quickly and easily determined. In other classes where the nonmetallic elements are dominant, it is less useful; however, it is always of value in revealing impurities, and, in a negative way, in showing what elements are absent. A discussion of the applicability of the instrument to the main mineral classes is given in the following pages.

Oxides. Since the element oxygen is a gas, it cannot be spectroscopically determined by burning minerals in the arc; however, since it forms compounds with most of the metallic elements its presence may usually be assumed. There are several different types of oxides, the most commonly occurring ones being the monoxides such as periclase (MgO), massicot (PbO), cuprite (Cu_2O), and zincite (ZnO), and the dioxides such as pyrolusite (MnO_2), cassiterite (SnO_2), rutile (TiO_2), baddeleyite (ZrO_2), and quartz (SiO_2).

Quartz, a form of silica is the most abundant of all minerals and is estimated to constitute 59 per cent of all rocks. Its oxygen does not produce an arc spectrum, and the scattered lines of silicon are all weak in the visible range. This is fortunate in mineral analysis, for, in spite of its high concentration in most rocks, its spectrum is unobtrusive. It is an excellent material on which to burn powdered samples for visual analysis. Most white sands, flints and cherts contain over 95 per cent of silica.

Other oxides are the sesquioxides with three atoms of oxygen and two of some metal in the molecule, as illustrated by arsenolite (As_2O_3), corundum (Al_2O_3) and hematite (Fe_2O_3). Still more complicated oxides are also numerous, as: spinel ($MgO \cdot Al_2O_3$), magnetite ($FeO \cdot Fe_2O_3$), chromite ($FeO \cdot Cr_2O_3$), chrysoberyl ($BeO \cdot Al_2O_3$), ilmenite ($FeO \cdot TiO_2$), and geikielite ($Mg \cdot Fe)O \cdot TiO_2$.

Alumina. Al_2O_3, forms over 15 per cent of the earth's crust, hence like silica it is a constituent of most rocks. It is the only oxide of aluminum. In the crystalline form, corundum, it has a hardness second only to the diamond. It commonly occurs, however, in combination with quartz and other minerals to form granites, diorites, clays, schists, gneisses, kyanites, etc. Pure alumina is a white powder used in making alum, mordants, pottery, etc. Aluminum may be spectroscopically determined by the violet doublet 3944-62, which, however, is too near the limits of visibility for easy observation.

Iron Oxides. Brilliant iron lines may be seen in the burning of almost any rock except white ones. Such iron minerals as hematite, limonite, magnetite and siderite will all register merely as iron in the spectroscope, but streak tests will distinguish them, since hematite drawn across an unglazed porcelain surface leaves a red mark, limonite a yellow mark, magnetite a black and siderite a white one. Red, yellow and black colors are characteristic of iron. Pyrite will betray itself by its explosive tendency in the arc and its sulfurous odor, marcasite by its color, pyrrhotite by its magnetism. Thus, in the classification of many minerals, it is necessary to apply auxiliary tests for hardness, density, streak, crystal form etc. to supplement the spectroscopic analysis.

The sesquioxide Fe_2O_3 (hematite), the ferrous oxide FeO, and the combination of the two Fe_3O_4 (magnetite) form the third largest mineral mass in the lithosphere and comprise nearly 7 per cent of it. Iron may be determined by any one of its numerous doublets, or by its other distinctive groups.

Silicates. Silicon may be detected by the faint visible line at 5948, or by much stronger ones in the ultraviolet. The silicates are more numerous than any of the other minerals, and of course all contain oxygen as well as silicon. Among the most widely spread of the silicates are the feldspars which also contain aluminum and either potassium, sodium or calcium; the pyroxenes, comprising magnesium, calcium, sodium and manganese silicates; the amphiboles, comprising magnesium, sodium, and calcium silicates, usually with iron; the sodalites, or sodium aluminum silicates; the garnets, comprising calcium, magnesium, iron and chromium silicates usually with aluminum; the epidotes, or calcium, aluminum and hydrogen silicates; and many others. The feldspars and garnets may usually be differentiated by their determinative metals.

Hydrous silicates, still more complicated in structure, are represented by the zeolites, the micas, the talcs and the kaolins. Of the micas, muscovite with its potassium, phlogophite with both potassium and magnesium, biotite with potassium, magnesium and iron, lepidolite with potassium and lithium, and zinnwaldite with lithium and iron may be differentiated by spectroscopic tests.

Calcareous Minerals, instantly recognized spectroscopically, come next in importance, with lime, CaO, forming about 5.1 per cent of the average rock mass. The limestones are formed in a variety of ways: by chemical precipitation when lakes or inland seas dry up, by the accumulation of the shells of marine animals, or by the cooling of hot springs. When these deposits crystallize with age and pressure and take on carbon dioxide, they become calcite or marble. These calcium deposits are both numerous and important, for they supply the materials for cement, mortar, statuary, etc. The spectrum of the average rock is very likely to show the presence of calcium. Such polymorphs as calcite and aragonite are, of course, spectroscopically indistinguishable.

Dolomite, a calcium magnesium carbonate is found in great masses, some of them hundreds of miles in extent, and is an important source of magnesium. It contains 13 per cent of the metal and is widely distributed.

Magnesian Minerals. The presence of magnesium in rocks and minerals is as easily determined spectroscopically as that of calcium. Besides dolomite, $CaMg(Co_3)_2$ there are several other magnesian minerals. While the oxide, MgO, is a common chemical compound and a common constituent in rocks, of which, on the average, it forms about 3.5 per cent, the mineral form of MgO, periclase, is rare. More familiar are epsomite, magnesite (which forms white massive veins rather than crystals), brucite, and serpentine of which there are inexhaustible deposits on the Pacific coast. The separation of the magnesium from the silica, however, presents the same practical difficulties as does that of separating aluminum from silica in clays. Most common rocks show well defined magnesium lines in their spectra.

Sodium Minerals. The first spectral lines one learns to recognize are those of the sodium doublet, 5890-96. Sodium oxide Na_2O, although a constituent of the lithosphere to about 3.7 per cent does not occur in nature as an independent mineral. It is always found in combination with other elements, but is so widely distributed that nearly all rocks will show sodium lines. The soda feldspars will have such a high percent of sodium as to cause reversal of the lines. The same effect will be observed in sodalite, Chile saltpeter, halite, albite, lazulite, borax, and thenardite.

Potash Minerals. Potassium oxide, K_2O, constitutes over 3 per cent of the rock mass of the earth, but occurs only in combination with other minerals. Spectroscopically, potassium has doublets at both ends of the visible spectrum and so is rather easily determined. Its lines soon become familiar in rock analysis, but they are less frequently seen than those of its chemical relative, sodium. Strong spectral lines may be observed in such minerals as the potassium feldspars (orthoclase), leucite, muscovite mica, sylvite, niter, and alum. It also occurs as a constituent of carnotite, the vanadium-uranium ore of Colorado.

Titanium Minerals. Titanium is present in a surprising number of rocks, and its spectral lines are so sensitive that even traces are easily detected. Rutile, TiO_2, constitutes a bit more than 1 per cent of the earth's crust and is especially concentrated in clays and shales. The following are among the most important of its two score minerals: ilmenite $FeO \cdot TiO_2$, rutile TiO_2, titanite $CaO \cdot TiO_2 SiO_2$, and perovskite $CaTiO_3$. Benitoite is a blue gem found only in San Benito county, California. Titanium is used as white pigment in pottery and to increase the tensile strength of steel.

Carbonates. Carbon cannot, of course, be determined in the carbon arc, since it is already there. Many important ores are in the form of carbonates, as illustrated by the following list:

Azurite $2CuCO_3Cu(OH)_2$
Calcite $CaCO_3$
Cerussite $PbCO_3$
Dolomite $CaCO_3MgCO_3$
Magnesite $MgCO_3$
Malachite $CuCO_3Cu(OH)_2$

Phosgenite $PbCO_3PbCl_2$
Rhodocrosite $MnCO_3$
Siderite $FeCO_3$
Smithsonite $ZnCO_3$
Strontianite $SrCO_3$
Witherite $BaCO_3$

Non-metallic Minerals. The spectroscope is not of much assistance in the determination of phosphates, arsenates, antimoniates, borates, tellurates, selenites, and sulfates so far as visual examination goes since their radicals yield no strong visible spectral lines. Selenium and sulfur even lack ultraviolet arc lines. The others have lines of moderate intensity. However, arsenic and sulfur give off such acrid fumes when burned in the arc, that they may be detected by their odors.

The haloid minerals, comprising the chlorides, bromides, iodides, and fluorides are spectroscopically indistinguishable when their metallic elements are the same, unless discharge tubes and special apparatus are used.

Rare Earths. The rare earths occur only in rocks of high sodium content, such as the allanites and the monazites, and many of them are usually found in the same ore. Some of them are not found in sufficient natural concentration to be detected even spectroscopically, but others form rather rich ores.

Rocks. Such white rocks as barite, brucite, celestite, dolomite, gypsum, cerussite, spodumene, wollastonite and albite all look about alike, but their spectra, due to different metallic elements, show them to be quite dissimilar.

Trace Elements. Vanadium and titanium traces are very common in widely scattered rocks, but their ores are scarce. Zirconium frequently occurs in granite, but its lines are weak and seldom seen. Traces of nickel and cobalt are also sometimes found in granite and in peridotite, but seldom in quantities or concentrations of commercial importance.

The following tables of crystal species and of mineral hardness are helpful in classification. Hardness runs from the very soft minerals, talc, calomel and graphite, hardness 1, to the diamond with hardness 10.

CRYSTAL FORMS

Isometric
Cubes: galena, pyrite, fluorite, halite, sylvite, leucite.
Octahedrons: magnetite, chromite, uraninite, diamond.
Dodecahedrons: garnet, cuprite, sodalite.
Trapezohedrons: garnet, leucite, analcite.
Pyritohedrons: pyrite, cobaltite, hauerite.
Tetrahedrons: tetrahedrite, sphalerite, helvite, diamond.

Tetragonal
Square pyramids: zircon, wulfenite, xenotime, vesuvianite.
Square prisms: zircon, scapolite, topaz, andalusite.

Hexagonal
Hexagonal prisms: beryl, apatite, vanadinite, quartz, calcite.
Tabular hexagons: graphite, ilmenite, pyrrhotite, hematite.
Hexagonal pyramids: apatite, corundum, quartz, chalcocite.
Trigonal prisms: tourmaline
Rhombohedrons: calcite, dolomite, alunite, siderite.
Scalenohedrons: calcite, proustite, barite, magnesite.

Orthorhombic
Prisms: stibnite, barite, celestite, pyroxene, orthoclase.
Tabular: barite, cerussite, calamine, wollastonite, albite.
Needlelike: stibnite, millerite, natrolite, zeolites.

SCALE OF HARDNESS

(1) Talc, calomel, graphite, molybdenite pyrophylite
(2) Gypsum, muscovite, kaolinite, borax, stibnite, autunite, sylvite
(3) Calcite, bornite, vanadinite, enargite, millerite, witherite
(4) Fluorite, xenotime, zincite, phillipsite, libethenite, manganite
(5) Apatite, cancrinite, hausmanite, dioptase, smithsonite, ilmenite
(6) Orthoclase, pyrite, magnetite, helvite, rutile, cassiterite
(7) Quartz, boracite, tourmaline, cordierite, danburite, staurolite
(8) Topaz, spinel, beryl
(9) Corundum, and the corundum gems: sapphire, ruby, topaz, emerald
(10) Diamond

Chapter VI

CHARACTERISTIC LINES OF THE ELEMENTS

Among the numerous spectral lines of each element there are almost always a few by which it may be easily identified, regardless of the complexity of the mineral, the alloy, or the compound. In the following pages an attempt has been made to point out these characteristic lines. Some elements have such brilliant and distinctive groups that recognition is almost instant and unmistakable; others have such widely scattered lines as to be identified only with difficulty, especially when the concentration is small, and the sample complex. In such cases one may have to place a single line within the line-group of another element known to be present, or measure its proximity to the line of an element introduced into the arc for that purpose, or select a solitary line in a relatively open field. When important lines are blocked out by the superimposed lines of another element, it is sometimes necessary to consider the element's weaker lines.

The elements are arranged alphabetically with their atomic weights, numbers, and symbols appended. Line intensities on the scale 1 to 10,000 are given in parentheses, since their importance is second only to line positions. Wave-lengths are expressed in Ångström units, carried to one decimal place. Arc lines are given unless otherwise indicated. Lines below 3900 are in the ultraviolet region. Neighboring lines of other elements are often given since they serve as spotters for those of the elements sought. Other information, either interesting or useful, has also been given.

Symbols

P—Persistent line
R—Reversed line
H—Hazy or wide

ACTINIUM AC (AT. WT. 227 AT. NO. 89)

    4180.1 Xe (500) Discharge tube
(60) Ac 4179.9   Discharge tube

 The actinium spectrum contains but few lines, one of which is given above. This line has an intensity of 60 on a scale of 10,000 and so is a pale line as seen in the glow of a Geissler tube. Its most important neighbor is the xenon line less than an angstrom away. Actinium is one of the half-dozen radioactive elements near the end of the atomic series, and is one of the most unstable.

 It was discovered by Debierné in 1899 while carrying out experiments in the Curie laboratory. Like radium it occurs in pitchblende. Actinium originates from one of the isotopes of uranium and passes in succession through a series of changes accompanied by the emission of alpha and beta particles. It passes so rapidly through one of these stages that half of its atoms change to the next lower form in less than four seconds! The half-life of the other stages varies from a few minutes to as long as twenty years. The end product of the disintegration chain is, like that of radium and thorium, lead.

 Although the spectroscope is the most delicate instrument known for detecting traces of nearly all elements, it is less sensitive than the Geiger-Müller counter in the detection of radioactive ones. However, although it is so sensitive to this group of elements, the counter cannot distinguish one from the other.

ALUMINUM AL (AT. WT. 26.97, AT. NO. 13)

    3968.4 Ca(500)   P(1000) Al 3092.7
P(3000) Al 3961.4   P(800) Al 3082.1
P(2000) Al 3944.0   R(200) Al 2575.1
    3933.6 Ca(600)   R(200) Al 2567.9

 Only the six lines given above are of value in spectroscopic analysis for aluminum. Two of these, although very strong, are close enough to the ultraviolet limit of visibility to reduce their value for visual use, especially in quantitative estimates. Together with the calcium lines given which so often show up in the same rocks, they make a very distinctive group. All four lines come out strongly on the photographic plate. Lines Al-2575-2567 are used in the quantitative analysis of magnesium alloys.

Aluminum is the most abundant of the metals and next in abundance to oxygen and silicon among all the elements. Its oxide, alumina, Al_2O_3, forms a part of nearly all rocks and when in crystalline form yields many of the important gems: blue sapphire, red oriental ruby, oriental topaz (yellow), oriental emerald (green), and oriental amethyst (purple). Corundum and emery are impure or granular forms of alumina. Artificial sapphires may be made in a few seconds by placing a little heap of aluminum oxide mixed with cobalt acetate upon a quartz block beneath the points of carbons arranged horizontally to form an arc. After fusing, the soft white and pink powders become a hard blue transparent mass that will scratch glass like a diamond.

The metal is malleable and ductile; it forms alloys with other metals such as magnesium and copper, but is also used pure for many articles. Its specific gravity is 2.71. Although more prevalent in the earth than iron, it is more difficult to extract from its ores and so more costly. It has wide use in the manufacture of airplanes, engine motors and machine parts. Bauxite and cryolite are its principal ores, but there have been promising recent attempts to extract it commercially from clays and feldspars. Cryolite is imported from Iceland, bauxite from South America. There are also large deposits of an inferior grade of bauxite in Arkansas. Aluminum is always present in granites, micas, schists, and clays, but nearly or quite lacking in peridotites, sandstones and limestones.

The war-time price of aluminum was $0.15 per pound.

ANTIMONY SB (AT. WT. 121.76, AT. NO. 51)

(20) Sb 6079.5   (300) Sb 2528.5
(20) Sb 6073.9   P(300) Sb 2175.5
      6063.1 Ba (200)   P(300) Sb 2068.3
(70) Sb 4033.5

The strong and persistent lines of antimony are all in the ultraviolet region, as represented by the right-hand group above. The visible lines do not ordinarily come out in the 110-volt arc, but the dense clouds of non acrid, white fumes, which rise when antimony ores are burned, betray its presence, as also does the white sublimate formed on the carbons near the points. When a bit of stibnite is burned in the arc upon a fragment of gypsum, a red and orange spot appears on the gypsum, surrounded by a yellow fringe.

Antimony has been known since the middle ages. Roasting the ore drives off the sulfur and leaves a white oxide, which, if it is mixed with carbon and roasted again, results in its separation. It is white, hard, brittle, of crystalline structure, a poor conductor of electricity and little affected by air. It forms numerous chlorides, bromides, and iodides.

The chief ore is stibnite, the trisulfide, imported from China and Japan, although also found in the western United States. The alloys of antimony expand upon cooling and so are used in casting type. Babbitt, an alloy of about 1 part copper, 2 parts antimony, and 5 parts tin, is used for machine bearings.

The price of antimony has varied in recent years from $0.06 to $0.16 per pound.

Argon A (At. Wt. 39.94, At. No. 18)

P(5000) A 8115.3 Discharge tube   (500) A 4806.0 Geissler tube
  (700) A 7503.8 Discharge tube   (300) A 4426.0 Geissler tube
           7488.8 Ne (500)   (500) A 4348.1 Geissler tube

The strongest argon line is just beyond the infrared limit of visibility; the second is near the same border. The spectral lines are grouped mainly in the red and blue regions of the spectrum. Argon, one of the inert gases, constitutes about 1 per cent of the air. Its presence was discovered by Cavendish, but it was another hundred years before two other Englishmen, Rayleigh and Ramsey, succeeded in separating it from the other gases of the atmosphere and in determining its chemical properties. Argon-

Characteristic Lines of the Elements 101

filled electric bulbs give a feeble violet light which excites fluorescence.

The price of argon is $7.50 per 2 cu. ft.

Arsenic As (At. Wt. 74.9, At. No. 33)

(200) As 5651.5 Spark   2354.8 Sn(150)R
(200) As 5585.3 Spark   P(250) As 2349.8
(200) As 5476.9 Spark   P(250) As 2288.1

Arsenic has no visible arc lines and only a few in the spark spectrum. Its ultraviolet arc lines, however, are fairly strong and readily photographed. It yields a garlic odor when burned in the arc. Fumes of the ore are sometimes rendered still more acrid by the sulfur which often accompanies it. A white sublimate sometimes forms on the carbons farther back from the tips than that of antimony.

Arsenic may function as a non-metal; it combines with the halogens and sulfur and with iron, zinc, and many other metals. The free element is gray, crystalline, and has a bright luster which quickly tarnishes. Its soluble compounds and fumes are all very poisonous; the room in which they are handled should be well ventilated. Arsenic is used in weed and pest poisons, in fireworks and in flint glass. Its chief ores are realgar and arsenopyrite.

The price of arsenic is now $1.85 per pound.

Barium Ba (At. Wt. 137.36, At. No. 56)

(2000) Ba 6141.7   PR(1000) Ba 5535.5
       6137.6 Fe (100)   PR (400) Ba 4934.0
       6136.6 Fe (100)   PR(1000) Ba 4554.0

The strongest barium line at 6141, may be instantly spotted in iron-bearing rocks by its 4-angstrom spacing from the iron doublet given above. Strong though it is, this line will not appear quite so quickly as the less brilliant but more persistent line 4554, as the concentration of barium in the sample is increased. Ba-5535 and Ba-4934 are also very sensitive. Barium penetrates the carbon tips of the arc and contaminates them so that it is often necessary to break or grind them off to clear the arc for the next analysis. Although barium constitutes but a small fraction of one percent of the earth's crust, it is so widely distributed that it will be found in a great many rocks.

Barium, strontium and calcium all have two electrons in their outer shells and their spectra are correspondingly similar. All have their strongest lines visible and disseminated over all the colors in about equal numbers. The groupings are so different, however, that there is no confusion of identity. The calcium lines form the most compact groups; the strontium lines run to groups of three; the barium lines are the most scattered.

The element closely resembles calcium; it decomposes water; powdered barium ignites spontaneously in air. The sulfate barite, when in crystal form looks like calcite, but is so much heavier that it was early called *heavy spar*. Witherite, the carbonate, is the next most common ore. Barium is used in white and phosphorescent paints; barium nitrate makes the green stars of fireworks. Its compounds are also used in sugar refining and in making glazed paper and cards.

Crude barite brings about $9 per ton.

BERYLLIUM (AT. WT. 9.02, AT. NO. 4)

	4578.5 Ca	(80)		(15) Be 6981.0	
	4573.8 Ba	(50)		6978.4 Cr	(125)H
	4572.7 Mn	(20)			
(15) Be	4572.6		(1000)	Be 3321.3	
	4571.9 Ti	(150)	P(2000)	Be 2348.6	
	4571.6 Cr	(50)			

The visible beryllium lines are few and weak, but Be-4572 is easily seen in beryl, the only ore of the element which is of commercial importance. Since the red line fails to show up in the 110-volt arc, the single blue line must be carefully checked. First check Mn-4754 to be sure there is no manganese in the arc; then check Ba-4554 to be sure there is no barium in the arc; Ti-4758-9 and Cr-5206 should also be checked to be sure of the absence of these elements. Finally, unless one is absolutely sure of the scale, copper should be added to the beryl in the arc and the position of its somewhat fuzzy line, Cu-4587, noted. If the distance between it and the beryllium line is 14 angstroms, or double the distance which separate the lines of the sodium doublet, one can be certain enough of the identity of the line. Of course, the stronger lines in the ultraviolet region can be photographed to show the presence of the element more easily.

Beryllium is a hard silver-white metal extracted from beryl, of which it forms about 6 per cent. It also forms about 23 per cent of phenacite, a beryllium orthosilicate which resembles quartz, but is a half-point harder. Beryl crystals are hexagonal, with a green tint due to chromium, and sometimes are of immense size, even weighing tons. Such crystals have been found in Maine and in the pegmatite veins of the Black Hills. The writer once searched the desert mountains of Nevada where the crystals have been reported, but without success.

The metal imparts its hardness to its alloys, 5 per cent in copper giving an alloy with 500 times the hardness of copper.

The war-time price of beryllium, due to the demand for it in manufacturing machine bearings, was from $100 to $300 per pound.

Bismuth Bi (At. Wt. 209, At. No. 83)

	4810.5 Zn (400) H		554.8 Fe (100)
(1000) Bi	4722.5	H (500)	Bi 5552.3
	4722.1 Zn (400)		
	4680.1 Zn (300)	PRH (3000)	Bi 3067.7
		PRH (500)	Bi 2897.9

Bismuth has two strong visible lines, the strongest less than half an angstrom from the zinc line 4722. In the absence of zinc, which may easily be checked by Zn-4680, 4810, bismuth is easily found. The weaker line Bi-5552 has a convenient iron spotter Fe-5554. Stronger lines occur in the ultraviolet region as indicated.

Because of the ease with which it is roasted from its ores, bismuth was known as far back as 1450, and perhaps even in ancient times. The metal has a reddish tinge, melts at 271°C and is a poor conductor of heat and electricity. One of its alloys, Wood's metal, consisting of Bi 4 parts, Pb 2 parts, Sn 1 part, and Cd 1 part, melts in hot water at 60°C, although that temperature is far below the melting point of any one of the metals alone and over 200°C below their average. The alloy is used to make plugs for sprinkler systems; it melts in case of fire and releases the water. A similar alloy of bismuth is used for steam-boiler plugs, the spectroscope being used to detect impurities that might affect the melting point and endanger the boilers.

Bismuth has still other strange properties. In contrast to iron and nickel, it is repelled by a magnet, and in contrast to most metals, it expands like ice when it congeals. Because of the latter property it is used in castings and stereotype plates. It is also used in thermocouples, resistant bronzes, and in medicine.

It is usually reclaimed from copper and lead ores, but is sometimes derived from bismutite in various parts of Europe. Deposits of bismutite are also found in the Carolinas and in Arizona.

The pre-war price of bismuth was $1.00 per pound.

BORON B (AT. WT. 10.82, AT. NO. 5)

P(500) B 2497.7
P(300) B 2496.7

Boron has fewer than 100 spectral lines of any intensity and none are visible. The ultraviolet doublet given above comprises its strongest and most persistent lines. At 220 volt some compounds show bands at 4744, 4747 and 4342. Boron compounds often give flames an intense yellowish green tint, but, since there are no green lines in its spectrum these must be due to molecular, rather than atomic activity. By moistening a borate with a drop of sulfuric acid, adding a couple of drops of alcohol, and touching with a match, one may produce the characteristic greenish flame. Early prospectors used such flame tests in locating the great desert deposits of borax in southern California.

Boron is a gray solid which melts only at temperatures above 2000°C. Compounds with aluminum, called *boron diamonds*,

Characteristic Lines of the Elements

have been artificially produced and are harder than the ruby. Borax and boric acid are its chief commercial forms.

The price of boric acid was $0.25 per pound in 1936; now it is not available; the price of boron $2.25 for 25 g.

BROMINE BR. (AT. WT. 79.92, AT. NO. 35)

P(300) Br 4816.7 Geissler Tube
P(400) Br 4785.5 Geissler Tube
P(250) Br 4704.8 Geissler Tube

Of the three visible lines listed above, all persistent, the weakest, Br-4704, is most persistent. Bromine has no arc spectrum. It is a caustic brown liquid which vaporizes at ordinary temperatures to a pungent gas. It is recovered from ocean brine and is used in the manufacture of dyes. It is also obtained from the salt wells of Ohio and Michigan. The reddish vapor was used as one of the poison gases in World War I. It is used in metallurgical operations and in various sedatives, overdoses of which are poisonous.

The price of refined bromine is $1.33 per pound.

CADMIUM CD (AT. WT. 112.41, AT. NO. 48)

```
              6439.0 Ca (150)
P(2000) Cd 6438.4                        4980.1 Zn (300)
                        H (200) Cd 4678.1
H(1000) Cd 5085.8
            5083. Fe (200)     (1000) Cd 3610.5
                             PR(1500) Cd 2288.0
```

In the absence of calcium, easily checked by the neighboring lines Ca-6462 and Ca-6449, cadmium is best determined by Cd-6438, its strongest and most famous line. If the calcium lines are present, Cd-5085, also a very strong line, may be used, with Fe-5083 serving as a spotter in spectroscopes having sufficient dispersion to separate lines 2 angstroms apart. Cd-6438 is a very strong singlet whose wave-length (6438.4696 A°) was measured by Michelson and later adopted as the international standard unit of length.

There are three ores of cadmium: greenockite, the sulfide; cadmium oxide; and otavite, the carbonate. Cadmium is also found in most zinc ores. The two strong lines, Cd-4678 and Zn-4680, are only 2 angstroms apart. In smelting, cadmium is reduced before zinc and so is easily separated from it. The oxide thus obtained is then mixed with carbon, and when it is reheated the cadmium distills over.

It is a lustrous metal capable of a high polish; it melts at 320°C and increases the fusibility of its alloys. Cadmium yellow, the sulfide, is one of the standard colors used by artists. An amalgam with tin is used for dental fillings; it is also used in electroplating and in low-melting alloys. The metal is now priced at $2.59 per pound.

CALCIUM Ca (At. Wt. 40.08, At. No. 20)

P(100) Ca 4455.8		PR(500) Ca 4226.7	
P(200) Ca 4454.7			
(100) Ca 4435.6		PR(500) Ca 3968.4	
(150) Ca 4434.9		3961.4 Al	(3000)P
(100) Ca 4425.4		3944.0 Al	(2000)P
4415.1 Fe	(600)	PR(600) Ca 3933.6	
4404.7 Fe	(1000)		
4383.5 Fe	(1000)	(100) Ca 3158.8	

There are calcium groups in all the color regions: red, between 6439 and 6500; orange, between 6100 and 6196; a yellow sextet between 5581 and 5603; blue between 4425 and 4455; and violet 3933-3968. Ca-4226, the most persistent of all calcium lines, is reversible and is the most valuable for making quantitative estimates. With most instruments, the blue group 4425-4455 will appear as a triplet and will form with the neighboring triplet Fe-4383-4404-4415 a very striking and symmetrical sextet from which the relative proportions of calcium and iron in the sample may be judged.

Calcium ranks fifth in abundance among the elements in the lithosphere. A majority of rocks will show a preponderance of calcium and iron lines, since together they form over 8.5 per cent of the earth, and both have far more numerous lines than the oxygen, silicon, and aluminum which outrank them in quantity. Since they are seen so often, the calcium groups soon become very familiar, but beside these there are strong individual lines which the analyst will look up a great many times before he will remember them, since single lines all look alike.

The most common calcium minerals are calcite, the carbonate, and gypsum, the sulfate, although apatite, the phosphate, and fluorite, the fluoride, are not infrequent. The burning of limestone, the carbonate, drives off carbon dioxide and leaves the oxide, lime, so important in plasters and cements. Quicklime, when wetted and mixed with sand, forms a silicate with it by a slow process; plaster of Paris, the product of calcined gypsum, hardens more quickly upon the addition of water. The phosphate is a valuable fertilizer.

The price of limestone is about $1.25 per ton.

Carbon C (At. Wt. 12.01, At. No. 6)

	C 6132.4 Band	C 4737.1 Band
	C 6004.9 Band	C 4216.0 Band
(100)	C 5354.1 Band	C 4180.8 Band
	C 5165.2 Band	P(400) C 2478.5 Arc

There are no visible carbon lines, but there are always numerous bands in the carbon arc. These are caused by carbon ions or compounds. In instruments of high dispersion these bands appear as groups of very fine lines crowded together. They are especially strong when other lines are absent from the arc and are often very weak or entirely absent when strong metallic vapors are present. The bright band heads or terminations listed above often assist in spotting spectral lines, but more often they merely clutter up the field and interfere with the identification of silicon and other weak lines. There is a strong, persistent carbon line in the ultraviolet region.

Carbon is present in all living bodies, alcohol, fuel oils and all of the thousands of synthetic organic compounds. Both diamond and graphite are pure carbon. The price of black powdered carbon is $0.65 per pound.

Cerium Ce (At. Wt. 140.13, At. No. 58)

	4528.6 Fe (600)	(100) Ce 5522.9
(30)	Ce 4528.4	(50) Ce 5353.5
(50)	Ce 4527.3 Ti (100)	(50) Ce 5409.2
	4526.9 Ca (100)	
	4526.1 La (100)	P(80) Ce 4186.6
(35)	Ce 4523.0	P(40) Ce 4165.6
	4522.8 Ti (100)	P(70) Ce 4040.7
	4522.3 La (200)	P(60) Ce 4012.3
(25)	Ce 4522.0	

Cerium has over 5000 charted lines—probably more than any other element—but none of these has even moderate intensity. All of its persistent lines are in the visible region. Cerium is probably best determined by 5522, one of its strongest lines. The 4522-4528 group, containing two close doublets, although made up of weaker lines, is more easily recognized because of its distinctive grouping. The comparative intensities of Ce-4527-4528 must be observed closely if iron and titanium are present.

Cerium, the best known of the group of metals termed the rare earths, was discovered independently in 1803 by two different scientists investigating Swedish minerals. Cerite, the silicate, is 56 per cent cerium. The samarskite of North Carolina and calciosamarskite of Canada are also rich sources of the element.

Cerium is used in spark-producing gas lighters, in sparking toy guns, and in gas mantles. Spectroscopically pure cerium oxide sometimes retails at $6.00 per gram, but cerium oxalate sufficiently pure for testing may be purchased at $0.65 per pound. The metal is $0.11 per gram.

Cesium Cs (At. Wt. 132.91, At. No. 55)

PR(5000) Cs 8521.1	PR(1000) Cs 4593.1
	4592.6 Fe (200)
	4591.3 Cr (200)
(500) Cs 6723.2	4556.1 Fe (150)
6717.6 Ca	4555.4 Ti (150)
6707.8 Li (3000)	PR(2000) Cs 4555.3

The strongest and most persistent cesium line Cs-8521 is in the infrared region and therefore can be detected only by using

Characteristic Lines of the Elements

special halite prisms and a special film, sensitive to this region; however, there are also very strong persistent visible lines, as shown. If the lines of titanium and iron are too strong, Cs-6723 may sometimes be preferable to the stronger lines, since it is clear of ordinary interference and close enough to the omnipresent lithium line 6707 to be easily spotted.

Cesium was the first element to be discovered by Bunsen and Kirchoff with their newly invented spectroscope in 1860. It is a soft silvery metal resembling rubidium and potassium and is the most electro-positive of all metals. It is used as a *getter* in radio tubes, to take out the last trace of oxygen, and is a valuable catalyst. It is often found in beryl, pollucite, and lepidolite.

Cesium chloride has been sold at $20 per gram, the metal at $4.00 per gram.

CHLORINE CL (AT. WT. 35.46, AT. NO. 17)

P(200) Cl 4819.4 Discharge tube
P(200) Cl 4810.0 Discharge tube
P(250) Cl 4794.5 Discharge tube

The strong and persistent chlorine lines are all blue. Chlorine is a greenish-yellow, irritating and noxious gas. It liquefies at 15°C, under a pressure of four atmospheres, into a yellow liquid. Its most familiar compound is sodium chloride, common table salt. Chlorine is a powerful bleaching agent and disinfectant; it is used in the purification of drinking water. Ocean salts contain about 30.5 per cent of sodium and 55 per cent of chlorine. The 1935 price of chlorine was $1.50 for 5 cu. ft.

CHROMIUM CR (AT. WT. 52.06, AT. NO. 24)

```
              5210.3 Ti (200)        PR(3000) Cr 4289.7
              5209.0 Ag (1500)R      PR(4000) Cr 4274.8
              5208.6 Fe (200)                 4260.4 Fe (400)
PR(500) Cr 5208.4                             4256.0 Ti (80)
              5207.8 Ti (25)                  4257.6 Mn (100)
PR(500) Cr 5206.0 Ti (40)                     4258.3 Fe (60)
PR(400) Cr 5204.5 Fe (125)                    4258.5 Ti (70)
              5202.3 Fe (300)        PR(5000) Cr 4254.3
```

Chromium is always easily recognized by the characteristic

green triplet 5204-5206-5208, so persistent that it has been observed when the concentration of chromium is only 0.0015 per cent. Although iron and titanium lines are indistinguishably close, their unequal intensities and greater number serve to prevent confusion, except that mere traces of chromium are sometimes obscured by brilliant iron and titanium lines. The lines of the 4254-4289 group are still more brilliant but also in a crowded field that makes them difficult to pick out, especially since they are more scattered. The green triplet is easily recognized, even in a small instrument, but under such conditions it will appear as a single, broad, bright line.

Chromium was discovered in 1797 by the French chemist, Vauquelin, who shortly thereafter also discovered beryllium. Chromium is a very hard crystalline metal. Chrome is derived from the Greek word for color and chromium compounds are generally highly colored, the chromates being yellow, orange or red, and the oxides green. Some are used as paint pigments, others give a green tint to glass. Other compounds are violet, blue, or black. The green color of beryl and of serpentine is due to traces of chromium.

Although a dozen minerals contain chromium in combination with magnesium, calcium, copper, arsenic, or lead, the only one to take rank as an ore is chromite, which is 25 per cent iron, 50 per cent chromium and 30 per cent oxygen. Chromium is used in varying proportions in steel to give hardness, strength, elasticity, and non-corrosive qualities.

The wartime price of 50 per cent chromite was $40 per ton.

Cobalt Co (At. Wt. 58.94, At. No. 27)

	5354.1 C. Band		H(1000)	Co 4813.4	
(500)	Co 5353.4			4810.5	Zn(400)H
(500)	Co 5352.0		(1000)	Co 4581.5	
(80)	Co 5347.4		(1000)	Co 4530.9	
	5345.8 Cr	(300)R	(1000)	Co 4121.3	
(600)	Co 5343.3				
(800)	Co 5342.7		PR(2000)	Co 3465.8	
(300)	Co 5341.3		PR(3000)	Co 3453.5	
(100)	Co 5339.5		PR(2000)	Co 3405.1	

Characteristic Lines of the Elements

Cobalt has so many lines and line groups that it is difficult to make a choice, but the close green doublet 5352-3 with lines of equal intensity is probably the most easily and certainly spotted, and the near-by group 5339-47 is corroborative. For instruments of lower dispersion the blue group of which line Co-4813 is but three angstroms from Zn-4810 will be better for determination. The persistent cobalt lines are in the ultraviolet.

The term cobalt was applied by the ancient Greeks to certain of their blue pigments, and it is mentioned in the writings of the middle ages, but its elemental nature was not discovered until 1735. Cobalt is a member of the nickel-iron-cobalt group of magnetic elements, all of which have similar spectra, containing many bright lines and numerous close doublets, and similar chemical properties. Its salts are pink but it imparts a vivid blue tint to glass. It is used in blue glazes for pottery. Cobalt blue is the artists' standard shade for painting skies. When it is heated with zinc oxide it produces a green pigment. Ink made with cobalt chloride is at first invisible, but upon being gently heated becomes visible.

Some cobalt alloys are so hard that they are used for tooling tungsten implements. An alloy with aluminum and nickel formed magnets, many times stronger than those of steel, which were used by the thousand in war planes. Cobalt is often associated with nickel; its ores include smaltite $CoAs_2$ and cobaltite CoAsS. The chief ore deposits are in Europe and Canada, but cobalt is also found in the granite near Boulder Dam.

Gems: Cobalt is the coloring matter in many gems. Take a clean white piece of quartz and place upon it a mixture of aluminum oxide and cobalt acetate. Place the mixture beneath the points of horizontally disposed carbons and strike the arc. A few seconds of burning will produce miniature sapphires as blue as a cobalt sky. Exchange the aluminum oxide for aluminum potassium sulfate (alum), burn for a few seconds longer than before and new colors will appear, green (emerald), violet, and pink. For yellow and green gems, substitute potassium dichromate for the aluminum compound. These synthetic gems will have the hardness of oriental ones.

The price of cobalt metal is $1.50 for 100 g.

Columbium Cb (At. Wt. 92.91, At. No. 41)

(200)	Cb	5900.6				4079.8	Fe (80)
		5899.3	Ti (150)	P (500)	Cb	4079.7	
		5895.9	Na (5000)			4062.6	Cu (500) H
(150)	Cb	5350.7		P(1000)	Cb	4058.9	Mn (80)
(400)	Cb	5344.1					Fe (80)
(150)	Ta	5341.0	Fe (200) Mn				
		5339.9	Fe (200)				
		5328.0	Fe (400)				

Probably the best test line for columbium is Cb-5900, fairly strong and less than 5 angstroms from the sodium doublet. The presence of titanium should be carefully checked because the Ti-5899 line may interfere. Two strong persistent lines in the violet region, Cb-4058-4079 have interfering iron and manganese lines, but there is a good copper spotter about half way between. The group Cb-5328-5344 contains columbium, tantalum, iron and manganese lines, and all of these metals are likely to be associated in columbite, the chief ore of the metal.

Columbium was discovered in a specimen of columbite sent to England by Governor Winthrop of Connecticut in colonial times. The discovery was not made, however, until Hacket examined the ore a century later, 1801.

Columbium is a steel-gray, hard metal whose alloys are said to have such great resistance to heat that they are preferred for jet-propelled craft. Ross rediscovered the element in 1846, renamed it Niobium, and this name is sometimes still used. The chief ores are columbite and pyrochlore; the former contains also cerium and the latter tantalum. (Coarse crystalline veins of feldspar and other minerals, called pegmatites, are the chief source of the ores. They occur in Europe, New England, and the Black Hills of South Dakota.)

In 1939 columbium sold at the rate of 5 g for $2.50.

Characteristic Lines of the Elements

COPPER CU (AT. WT. 63.57, AT. NO. 29)

```
            5227.1  Fe (400)                 5785.9  Ti (100)H
P(700) Cu  5218.2                  (1000) Cu 5782.1
            5216.2  Fe (300)                 5777.6  Ba (500)R
            5209.0  Ag (1500)
P(600) Cu  5153.2  Na (600)       PR(3000) Cu 3273.9
P(500) Cu  5105.5                 PR(5000) Cu 3247.5
```

The strongest visible copper line Cu-5782 lies between titanium and barium lines, and also between mercury lines, about 100 angstroms from the sodium doublet. It is not persistent, however, and when only traces of the element are present, it will fade out before the somewhat weaker green triplet Cu-5105-5153-5218, by which the metal is most easily identified in smaller instruments. Still stronger persistent lines are found in the ultraviolet region. Copper can be detected in amounts of less than 0.0015 per cent of the sample being analyzed.

Copper was one of the first metals used by man; it is mentioned in the Bible and by Homer, and implements of the metal have been found in excavations of ancient cities. The American Indians used the native copper found near Lake Superior. The alchemists of the middle ages believed that the copper precipitated by iron constituted a transmutation from one element to the other.

Copper, silver and gold are of the same chemical family and all have comparatively simple spectra with some fairly strong lines. They are all chemically sluggish and are sometimes found in their native state—gold usually so. Copper is a red metal, very ductile, and subject to tarnishing in moist air. Next to silver, it is the best conductor of electricity.

The chief copper ores contain either sulfur (chalcopyrite, chalcocite, and copper pyrite), or carbon and oxygen (malachite and azurite). The carbonates, mixed with a flux and coal, are roasted; the recovery of copper from the sulfide ores requires more elaborate treatment.

Copper is used in coins, drawn into wire for electric uses, alloyed with zinc in brass, with tin in bronze, with zinc and nickel for the white alloy, German silver, and with aluminum for a gold-colored bronze.

Electrolytic copper cost $0.12 per pound in 1943.

Dysprosium Dy (At. Wt. 162.46, At. No. 66)

	4957.6	Fe	(300)	P(200) Dy 4211.7 Tb (25) Ti (30)
(20) Dy	4957.3			P(150) Dy 4077.9 Tb (25) La (600)
(30) Dy	4731.8	Tb	(4)	P(150) Dy 4045.9 Tb (25) Fe (400)
	4728.4	La	(400)	P(400) Dy 4000.4 Tb (15) Fe (35)
(70) Dy	4589.3	Tb	(15)	

The persistent lines of dysprosium are in the crowded violet region and the lines of terbium shadow those of dysprosium. Frequently, the lines of the two elements can be distinguished only by their intensities. The lines of iron, titanium, and lanthanum further complicate the determination. The green region is not so crowded but the dysprosium lines here are weak. In spite of its more than 2000 lines, the identification of dysprosium is still difficult because most of the lines are of such low intensity: 2, 4, 6, and 8—and the strongest lines are all scattered and present little pattern.

Dysprosium was discovered in 1886 by Lecoq de Boisbaudran, who also discovered samarium and holmium, two more of the rare earth elements. The element has never been isolated, but has been recognized in a variety of minerals, e.g., blomstrandine, euxonite, fergusonite, gadolinite, polycrase, and xeontime. Its salts, like those of erbium, holmium, and thulium are highly colored. All have characteristic absorption spectra. For a complete list of the rare earths see erbium, the next element.

Erbium Er (At. Wt. 167.2, At. No. 68)

	5424.0 Fe (400)	(20) Er 4087.6
(30) Er	5422.8	(35) Er 4007.9
(20) Er	5419.5	
	5415.2 Fe (500)	P(25) Er 3906.3
(50) Er	5414.6	P(20) Er 3692.6

Erbium, like dysprosium, has over 2000 lines, but they are still weaker than those of dysprosium. The yellow triplet Er-5414-5419-5422, containing its strongest line and interlaced with

Characteristic Lines of the Elements

the doublet Fe-5415-5424, is probably its most distinctive group.

Most of the rare-earth elements have their persistent lines in the violet region; the persistent lines of erbium, terbium, thulium, and ytterbium lie in the ultraviolet region.

TABLE 14

THE RARE EARTHS

Element	Atomic Number	Discovered by	Year of Discovery
Lanthanum	57	Mosander	1839
Cerium	58	Mosander	1839
Praseodymium	59	Welsbach	1885
Neodymium	60	Welsbach	1885
Illinium	61	Hopkins	1926
Samarium	62	Boisbraudran	1875
Europium	63	Demarcay	1901
Gadolinium	64	Marignac	1886
Terbium	65	Mosander	1843
Dysprosium	66	Boisbaudran	1907
Holmium	67	Boisbaudran	1886
Erbium	68	Mosander	1843
Thulium	69	Cleve	1879
Ytterbium	70	Marignac	1878
Lutecium	71	Welsbach	1907

The brightest lanthanum lines and its densest line groups are in the violet region. This is also true of elements 57 to 63. Elements 64 to 71, except dysprosium, have their densest groups and persistent lines in the ultraviolet region, those of lutecium shifting to 3000 Å.

Erbium occurs in the same minerals and has the same characteristically colored salts as dysprosium.

The price of erbium nitrate is $0.40 per gram.

EUROPIUM EU (AT. WT. 152, AT. NO. 63)

```
              6021.7 Mn (80)          (1000) Eu 5577.1
(1000) Eu 6018.1                             5572.8 Fe (300)
              6016.6 Mn (80)          (1000) Eu 5570.3
              6013.4 Mn (100)                5569.6 Fe (300)

(300) Eu 5972.7 Ba (100)              (2000) Eu 5831.0
             5971.7 Ba (150)          (2000) Eu 5765.2
(2000) Eu 5967.1                      PR(150) Eu 4129.7
(1000) Eu 5966.0                      PR(200) Eu 4205.0
```

Europium has a greater number of very strong lines than any other element, and so is in striking contrast to the majority of the rare earths with only a multitude of weak ones. Moreover, the europium lines are scattered through the whole visible spectrum instead of being bunched on the ultraviolet border with the others. One of its brilliant lines, Eu-6018, may easily be spotted when manganese is present since it converts the red trio of that element into a quartet. One of its several doublets Eu-5966-5967 is 4 angstroms distant from a similar close doublet of barium; another doublet Eu-5570-5577 interlocks with one of iron.

Europium is one of the rare earth metals, so rare, in fact, as to be of academic interest only. In spite of its brilliant spectrum it was not discovered until 1901. It belongs to the terbium group, which also contains gadolinium. It is known only in the form of its salts. It occurs in the monazite sands of Brazil, but always in minute quantities. There is little chance of finding an ore rich enough to give its spectrum without preliminary concentration. Nevertheless, its spectrum has been discovered in the sun.

For a complete list of rare earths see Table 14.

Fluorine F (At. Wt. 19, At. No. 9)

(300) F 6632.7 Band	P(500) F 6902.5	Discharge Tube
(200) F 6064.4 Band	P(1000) F 6856.0	Discharge Tube
6062.3 Band		
6060.4 Band	(100) F 5296.8	Band head Arc
6058.6 Band	(150) F 5292.9	Band head Arc
	P(200) F 5291.0	Band head Arc

Fluorine yields spectral lines in the discharge tube but none in the arc. However, calcium fluoride, its most frequent mineral compound, betrays its presence by brilliant bands, both the strength and position of which vary with the voltage of the arc. Using 220 volts direct current, the most brilliant band head will be 6632, while with the 110-volt alternating current arc that band disappears and 6064 takes first rank. The orange band is very brilliant at the head and shades off toward the yellow; the green

Characteristic Lines of the Elements

band, less brilliant, also shades toward the yellow but in the opposite direction.

Fluorine was recognized as an element about a century before it was isolated as a greenish-yellow gas by Moissan in 1886. It has a disagreeable, irritating odor and is poisonous. It is very chemically active and belongs to the halogen group of elements. Fluorine combines with silicon and is used to etch glass. It unites explosively with hydrogen and ignites organic materials and such organic liquids as alcohol, ether, and turpentine. Hydrofluoric acid has many of the properties of the element and is about as toxic.

Its most common mineral is fluorite, many crystals of which fluoresce a brilliant blue. Lenses made of it have small dispersion. It is used as a flux in roasting refractory ores. Cryolite, topaz, and wavellite also contain fluorine.

A recent price of fluorite is about $30 per ton.

GADOLINIUM Gd (At. Wt. 156.9, At. No. 64)

(200) Gd 6857.1
(500) Gd 6846.6
        6842.6 Fe (30)
        6841.3 Fe (50)
(400) Gd 4865.0
(100) Gd 4862.6
(100) Gd 4861.8

                        4347.4 Hg (200)
(100) Gd 4347.3
 (50) Gd 4344.4
(200) Gd 4342.1
(200) Gd 4341.2

(500) Gd 4327.1
(500) Gd 4325.6

PH(200) Gd 3545.1

One of the best groups for the detection of gadolinium is the couplet 6846-6857 with a near-by iron doublet as a spotter. The blue quartet is a compact group with Hg-4347 as an excellent spotter. It is always best to keep the spectrum as simple as possible and to check with elements having but few lines. Mercury is especially adapted for spotting since it gives bright flashy lines and then quickly evaporates leaving the arc clear. It is easier to use a little calomel on top of the sample than to use the pure mercury. *Do not use a large quantity since mercury fumes are poisonous.* Two of the lines of the blue triplet, with a space of less than an angstrom between them, will blend unless one has a good spectroscope, but the strong line Gd-4865 will stand clear and define the group. The only strong persistent line is in the ultraviolet region.

Gadolinium was named in honor of Gadolin, a Finnish chemist who in 1794 investigated gadolinite from Sweden and showed that it contained a new element. Others have confirmed the discovery and gone further to show that the mineral contains half a score of the rare earths. Gadolinium, although more abundant than the other elements of the terbium group, has never been isolated, but the sesquioxide has been secured, and is offered at $0.08 per milligram.

GALLIUM GA (At. Wt. 69.7, At. No. 31)

4177.5 Fe (100)	4034.4 Mn (250)
4176.5 Fe (100)	4033.0 Mn (400)
4175.6 Fe (100)	PR(1000) Ga 4032.9 Fe (80)
4174.9 Fe (60)	4030.7 Mn (500)
4172.7 Fe (160)	4030.4 Fe (120)
PR(2000) Ga 4172.0 Fe (80)	

The strongest and most persistent gallium line Ga-4172, lies within a crowded iron group, but in any considerable concentration will stand out clearly from the weaker ones of iron. The second persistent gallium line Ga-4032 is in the midst of a strong manganese group, but since its usual association is with zinc rather than manganese these are not likely to be present. The ultraviolet lines are weak.

Gallium was discovered in 1875 by Boisbaudran and given the ancient name of his country. It was the first of four elements which he discovered with the spectroscope. A dozen or more new elements were discovered with the instrument in the quarter century following its application to mineral analysis by Kirchoff and Bunsen in 1861.

Gallium is a medium hard, somewhat malleable metal, not

tarnished in air, and with a melting point so low (86°F) that it may be liquefied by merely holding it in the hand. It may be whittled with a knife, and bent without breaking. It has many points of resemblance to aluminum and indium, which are in the same atomic group: low melting points, a valence of 3, and few bright spectral lines.

It is found sparingly in some zinc blends, in aluminum ores and in germanite, a South African mineral which also contains copper, iron, zinc, germanium and sulfur.

The price of gallium metal is $6.00 per gram.

GERMANIUM Ge (AT. WT. 72.6, AT. No. 32)

	4227.4	Fe (300)	(20)	Ge 4685.8
P(200) Ge	4226.5	Ca (500)R		4680.1 Zn (300)
	4224.1	Fe (200)		
	4222.2	Fe (200)	P(300)	Ge 3269.4
	4219.3	Fe (250)		
	4217.5	Fe (200)	P(1000)	Ge 3039.0
	4216.1	Fe (200)		

Germanium has few spectral lines. The strongest visible line Ge-4226 is but a fraction of an angstrom from calcium's most persistent line and within a group of iron lines of equal intensity with itself. Germanite, the chief ore, contains no calcium, and so its analysis is not complicated. Where calcium is present there are still stronger ultraviolet lines to fall back upon. In sufficient concentration the weak line 4685 may be spotted by the familiar zinc line 4680.

Germanium was discovered by Winkler in 1886, a year marked by the discovery of three other new elements, holmium, gadolinium, and fluorine—all by different investigators. The previous year Welsbach had discovered two rare earth elements, to make a total of five in three years.

It is interesting to note how the discovery of new facts by one man is likely to stimulate and touch off a whole chain of discoveries by others. Mendeleeff, the brilliant Russian chemist, prophesied that an element would eventually be found to fill the vacant space between silicon and tin in his periodic table. He lived to see this prediction fulfilled fifteen years later in germanium.

Germanium is a brittle, whitish metal found in some silver ores and in the African mineral, germanite, which may contain also iron, copper, sulfur, gallium and sometimes molybdenum. In a specimen examined by the writer there was also zinc, and Ga-4172 was at times brighter than Ge-4226. Germanium melts at 900°C and oxidizes under the influence of heat; it crystallizes in the octahedral system.

It may be purchased for about $3.50 per gram.

GOLD AU (AT. WT. 197.2, AT. NO. 79)

	6282.6	Co (300)H		4792.8	Co (600)
(700) Au	6278.1		H(200) Au	4792.6	
	6261.0	Ti (33)		4792.5	Cr (200)
				4791.2	Fe (200)
	5838.6	Cb (200)			
(400) Au	5837.3		PR(250) Au	2675.9	
	5826.2	Ba (150)H	PR(400) Au	2427.9	

Gold is most easily determined by its strongest line 6278, with Ti-6261 as a spotter. This is by far the best of its visible lines to use because of its strength and the absence in its vicinity of interfering lines with which it is likely to be confused. Au-5837 is also free from interference, except for the possibility of columbium. Au-4792 is unsatisfactory because of chromium, cobalt, and iron lines that would have to be carefully checked. Ultraviolet lines are not so strong but are more persistent.

If a little heap of powdered quartz is placed upon a fragment of quartz beneath the carbon points, a bit of gold alloy the size of a grain of wheat may be melted into it and held in the arc instead of forming a molten ball and rolling away. Gold is not evenly distributed throughout its ore, but is usually found only in minute particles here and there. Even in very profitable ore these particles may not happen to be in the arc at the time of inspection, and so may be missed in the test.

Gold was known to ancient Egypt, India, and Rome; it was used to adorn the temples of the Aztecs. Prehistoric mines have been found in Africa, Europe and America. Gold has always been precious and there has never been enough of it to satisfy the desires of man. The great ambition of the alchemists of the middle ages was to convert the baser metals into gold.

It is almost always found native; only aqua regia and cyanide will dissolve it. Its chief uses are for ornamentation and for stabilizing national monetary systems.

The present value of gold is $36.00 per ounce.

HAFNIUM HF (AT. WT. 178.6, AT. NO. 72)

(150) Hf 7131.8
(100) Hf 6980.9
       6978.4 Cr (125)H
(100) Hf 6818.9
(100) Hf 6644.6
       6643.6 Ni (300)

       5316.7 Co (300)
(100) Hf 5311.6
P(25) Hf 4093.1
P(80) Hf 3134.7
P(80) Hf 3072.8

The hafnium lines are so widely scattered that its spectrum has little pattern. The line Hf-7131 is too far toward the infrared region for convenient observation; Hf-6980 has a good chromium spotter; Hf-6818 is most favorably placed for observation in a thinly occupied field; Hf-6644 has a nickel, and Hf-5311 a cobalt spotter. All lines are rather weak and the persistent ones weaker still. Hafnium has, altogether, over 1500 feeble lines. The oxide gives long and complex bands. Its lines have been found in the solar spectrum.

Hafnium was discovered through Roentgen-ray spectroscopy by Coster and Hevesy in 1923, and was named after the Danish capital the place of its discovery. Its atomic number is established as 72 by the X-ray spectrum, therefore it immediately follows the rare earth elements in the atomic series. Hafnium was first discovered in zircon from Norway, and it seems to be an inseparable companion of zirconium, which it resembles in its chemical characteristics as well as in its spectrum. The ratio of Zr to Hf is always about 20 to 1. Hafnium has been united with potassium and fluorine to form a crystalline compound free from zirconium, and is then separated from these crystals to produce the free element. As in the case of germanium, its properties had been predicted by Mendeleeff. Its melting point is high and its emission of electrons so great when heated as to make it highly suitable for the cathodes of electron tubes.

HELIUM He (At. Wt. 4.003, At. No. 2)

P(1000) He 5875.6 Discharge Tube
P(300) He 4685.7 Discharge Tube
P(1000) He 3888.6 Discharge Tube

Helium has only 110 discharge-tube lines in its spectrum strong enough to be of any value in qualitative analysis, although some of its fainter ones may be of importance in showing its atomic structure or the completeness of the spectral series. Those given are the persistent ones. The usual number of spectral lines of the inert gases, of which helium is the lightest, is about 1200; helium has but one-tenth as many. All these gases have brilliant lines of (1000) or (2000) intensity, except radon, the last of the series.

Helium was spectroscopically discovered in the sun long before it was known on earth. Lockyer was the first to note He-5875 15 angstroms from the sodium doublet during the eclipse of 1868. Ramsay was the first to isolate it in 1895. In 1903 he observed that the accumulated emanations of uranium gave the same spectrum as helium.

Eventually, fractional evaporation showed that it was also present in the air. It is preferred to hydrogen for filling blimps and balloons since it is not inflammable. It is found in usable quantities only in the oil wells of Texas and other parts of the South. In medicine it is used in treating certain respiratory conditions.

Helium was liquefied as early as 1909 but was not solidified until its temperature was lowered to within 1°C of absolute zero and subjected to a pressure of 26 atmospheres.

Helium has been sold at less than $10.00 per 1000 cu. ft.

Characteristic Lines of the Elements 123

HOLMIUM Ho (AT. WT. 164.94, AT. NO. 67)

(200) Ho 5982.9
        5978.5 Ti (125)
(125) Ho 5973.5
        5972.7 Ba (100)

(200) Ho 5933.7
(200) Ho 5921.7

(100) Ho 5955.9
        5953.1 Ti (150)
(200) Ho 5948.0

(200) Ho 5882.9

P(200) Ho 3891.0

The holmium doublet Ho-5921-5933 has components of equal strength. The doublet Ho-5973-5982 (with Ti-5978 midway between) affords a good check. Ho-5882 forms a very evenly spaced triplet with the lines of the sodium doublet. Holmium's 800 significant lines are but one-third the number of those of the typical rare earth element, but they are of about average intensity as compared with the others. Only thulium and lutecium have fewer lines.

Holmium was discovered about 1878 by either Soret, Cleve, or Boisbaudran or independently by all three; there seems to be considerable difference of opinion. Holmberg secured the oxide in 1911, but the pure element has never been isolated. The rare earth elements are, of course, all metals, but as in the case of holmium, it has been so difficult to secure enough of their simple compounds with which to work that only their salts or oxides have been obtained. Since the final product in so many cases is but an earthy substance like alumina or magnesia, they are frequently called *rare earths*.

Holmium oxide is yellowish; the salt solutions show a strong absorption spectrum. For a complete list of the rare earths see erbium.

The oxide is priced at $1.50 per milligram.

HYDROGEN H (AT. WT. 1.008, AT. NO. 1)

    P(3000) H 6562.8 Discharge Tube
    P(500) H 4861.3 Discharge Tube
    (300)   H 4340.4 Discharge Tube
    (100)   H 4101.7 Discharge Tube

The strongest hydrogen line, H-6562, is a very brilliant red one which may be excited either by producing a high-voltage spark in the gas at atmospheric pressure or by passing an electric discharge through a Geissler tube filled with rarefied hydrogen. The passing of many electrons from the third level to the second is the cause of its brilliance.

Hydrogen has the simplest atomic structure of all the elements with a nucleus consisting of a single proton and a planetary system having but a single electron. As would be expected, it has accordingly, the simplest of atomic spectra, with but 21 lines of intensity higher than 1. Persistent H-6562 is the brightest visible spectral line emitted by any of the gases, and except for the infrared argon line A-8115 (5000), the brightest in any region.

Although hydrogen gas had been produced in the middle ages by treating metals with acids, its real nature was not understood until the experiments of Cavendish in 1766. Its presence in the sun and the stars is shown by the spectroscope.

Hydrogen and carbon unite in different proportions to form hundreds of hydrocarbons such as methane, paraffin, kerosene, gasoline and benzine. The gas unites explosively with chlorine in sunlight to form hydrochloric acid; it also unites explosively with oxygen to form water; it unites with sulfur to form an odorous gas, with nitrogen to form ammonia, and with bromine, iodine, and phosphorus. With carbon, it forms the main constituent of fuel gas. It issues from deep oil wells, usually combined with carbon as methane (CH_4) and is then termed *natural gas*. It often forms part of the gases issuing from volcanoes.

California natural gas comes at about $1.00 per 500 cu. ft.

ILLINIUM IL (AT. WT. 147, AT. NO. 61)
Between .33596 and .31311 X-ray *

The bright-line spectrum of illinium is unknown. The element was discovered by Hopkins of the University of Illinois in 1926;

---

* This reading was obtained with an X-ray spectrometer. In this instrument the diffraction grating consists of the faces of different crystals. The element to be analyzed is placed upon the anti-cathode within the X-ray tube, and the rays reflected from it are then diffracted by the crystal.

Characteristic Lines of the Elements 125

and with its discovery the list of rare earths became complete. In the same year Corke, James, and Fogg independently obtained the element and determined its X-ray spectrum. Subsequently two Italian investigators claimed prior discovery and proposed the name florentium. It occurs in such minerals as monazite and gadolinite. Still other workers have confirmed the existence of the element. A rare earth metal is seldom found alone; where one is found others may be expected. In one mineral one may predominate and in another mineral another, but it is believed that some of each rare earth will be found in any considerable deposit of an ore containing one of them. Even though chemists knew in what minerals the missing element No. 61 could be expected, yet it was present in such minute quantities that for many years the search for it was in vain.

Although the most valuable commercial deposits of monazite are in Brazil, it also occurs in Russia, Norway, India, Africa, Canada and the United States. In both Brazil and the United States monazite is found in the red and black sands formed by the decomposition of granites and gneisses and their pegmatite veins. The deposits are not very limited geographically, but for some of the elements they are infinitesimal in amount. Richer masses of cerium or lanthanum sometimes occur, but illinium and samarium have never been found in anything but the merest traces. Neither the metal nor the salts of illinium have yet been obtained in pure form.

INDIUM IN (AT. WT. 114.76, AT. NO. 49)

```
                4524.7 Sn  (500)        P(2000) In 4101.7
                4518.0 Ti  (100)
                4512.7 Ti  (100)        P(1500) In 3256.0
                4511.9 Cr  (80)
P(5000) In      4511.3 Sn  (200)  Ta (300)
                4509.3 Cu  (150)
```

In-4511 is so strong and persistent that the visible determination of indium is usually easy; but since tin has a moderately strong line on almost exactly the same wave-length it is necessary to check Sn-4524, but a few angstroms away, in order to ascertain which element is responsible for the line. If both lines are present and the intensity of Sn-In-4511 is equal to or greater than that of Sn-4524, then indium must be present as well as tin to account for the inverted intensities. The check is especially important since both elements are frequently present in the same ore. The other strong indium line In-4101 should also be checked for the possible presence of tantalum.

Indium is a silvery metal belonging to the trivalent group containing aluminum, gallium and thallium. Indium and gallium are both found in zinc ores, and sometimes also in association with lead, cadmium, and tin. It is soft, malleable and easily fused, and its two blue spectral lines have been observed even in the bunsen flame. It combines with oxygen either as a monoxide or as the sesquioxide In_2O_3. It has been in demand in recent years for use in cores for metal castings. After the outer metal has hardened, the indium is melted out to leave the casting hollow. It is used especially in aluminum castings since aluminum has a melting point of 660°C and indium 167°C. Indium is also used in atomic research to determine the slowing down of neutrons in graphite. It indicates this by the intensity of its activation, along with such other detectors as rhodium and iodine.

The price of the spectroscopically pure metal is $20.40 per gram.

IODINE I (AT. WT. 126.92, AT. NO. 53)

(900) I 5464.6 Discharge Tube
(900) I 2062.3 Discharge Tube

Like sulfur, iodine melts at less than 120°C without yielding an arc spectrum. The discharge tube, however, shows strong yellow and ultraviolet lines.

Iodine was discovered in kelp by Courtois in 1811 and for many years kelp was the chief source of the element. It was afterward obtained more easily from the nitrates of Chile, and still later even more cheaply from the brine of certain deep exploratory oil wells. It boils at 184°C giving off a dense violet vapor.

Iodine, one of the halogens, combines with sodium, potassium, and with most of the other metals, and with chlorine, oxygen, and sulfur, but its action is sluggish in comparison with that of other halides.

Iodine is used to produce color tints in some of the coal-tar dyes. The iodides, according to Chinese records, have been used in medicine since 2000 B.C. The latest application in this field is the use of radio-active iodine for diseases of the thyroid. A mild tincture of iodine is used locally on skin or mucous membrane as an antiseptic, iodoform is used as an antiseptic and deodorant. Iodine stains may be removed by washing first with alcohol or ammonia water, then with hot, soapy water. Large doses of iodides have usually been recommended for actinomycosis, but recently sulfadiazine and penicillin have appeared to be distinctly more valuable. Potassium iodide is often prescribed in asthma.

The price of crude iodine is now $1.92 per pound.

IRIDIUM Ir (AT. WT. 193.1, AT. No. 77)

(400) Ir 4426.2
       4404.7 Fe (1000)
(300) Ir 4403.7
       4401.3 Ni (1000)
(400) Ir 4399.4
       4396.3 Ag (100)
       4383.5 Fe (1000)

P(100) Ir 3513.6
P(100) Ir 3220.7
HP(25) Ir 2924.7
HP(40) Ir 2849.7

Iridium may be spotted by the strong doublet 4399-4403, with Ir-4426 not far away for a check. Adding silver to the arc will make the doublet appear as a triplet. The doublet also lies between the very strong iron lines 4383-4404. Although the most persistent lines are in the ultraviolet region, the strongest ones are visible.

Iridium was discovered in 1803 by Tennant, who found that a residue which remained after dissolving platinum with aqua regia consisted of two new metals, osmium and iridium. Iridium is very hard and brittle. It occurs in the mineral osmiridium usually accompanied by one or more others of the platinum metals. Iridium is found in Colombia, the Urals, Australia, and Brazil. It occurs in the black sands of California and Oregon which are washed for gold and platinum.

The rarity of the metal prevents wide application, but it is used to tip gold pens and in electrical coils designed to give standard resistance. Since the metal has minimum linear expansion it was used with platinum in the alloy of which the standard meter bar at Paris was made. An alloy consisting of 9 parts platinum and 1 part iridium is as elastic as steel and wholly untarnishable. In 1880 a method of electroplating with iridium was perfected; by this method it may be used as a protective coating for surgical instruments, balances, etc. The members of the platinum family typically have a great wealth of spectral lines, as do those of the iron-cobalt-nickel group.

The price of iridium is about five times that of gold.

IRON FE (AT. WT. 55.85, AT. NO. 26)

(100)	Fe 6137.6	(600)	Fe 4415.1
(100)	Fe 6136.6	(1000)	Fe 4404.7
			4401.5 Ni (1000)
(100)	Fe 6016.6		4399.4 Ir (400)
(100)	Fe 6013.5		4384.7 V (125)R
		(1000)	Fe 4383.5
(200)	Fe 5341.0		4379.2 V (200)R
(200)	Fe 5339.9	(1000)	Fe 4325.7
		(1000)	Fe 4307.9
(1000)	Fe 4925.2	(1000)	Fe 4271.7
PR(1000)	Fe 3719.9		

Over 5000 lines of iron have been plotted, and it probably has more lines of average intensity than any other element. A very distinctive visible group Fe-4383-4404-4415, mentioned in the discussion of calcium, forms with that element a symmetrical sextet observable even with small instruments. The group is important, moreover, because it contains easily recognizable lines of nickel, iridium, and vanadium. Iron is so omnipresent in rocks that its many singlets, doublets and larger groups soon become very familiar. Because of its wealth of lines, both visible and ultraviolet, its spectrum is generally used for locating the lines of unknown elements. At times, however, its host of lines may become very annoying, especially when they obliterate desired weaker ones. Although there are many intense visible lines the persistent ones are in the ultraviolet region.

Iron was known in ancient Assyria and in Egypt. It was employed to some extent by most ancient nations but its use did not become general until the spread of the Roman Empire. Pure iron is soft, ductile, and magnetic. It is used in greater quantities and for more varied purposes than any other metal. The chief ores are hematite, limonite, and magnetite. Brassy pyrite crystals are very frequent in rocks. Iron alloys each have special properties; carbon gives hardness, silicon softness, chromium proof against rust, manganese elasticity, tungsten toughness, titanium rigidity, sulfur and phosphorus brittleness, etc.

The price of pig iron has often been less than one cent per pound.

Krypton Kr (At. Wt. 83.7, At. No. 36)

P(3000) Kr 5870 Discharge Tube
P(2000) Kr 5570 Discharge Tube

Krypton was first identified as a new element by the very strong, yellowish green lines in the discharge tube glow. It is one of the 6 inert gases that make up air, but it forms only one part in millions of the volume of air. Its spectrum contains over 1000 lines.

TABLE 15
The Inert Gases

Element	Atomic Number	Discovered by	Year of Discovery
Helium	2	Ramsay	1895
Neon	10	Ramsay and Travers	1898
Argon	18	Ramsay and Rayleigh	1894
Krypton	36	Ramsay and Travers	1898
Xenon	54	Ramsay and Travers	1898
Radon	86	Curie and Curie	1900

None of these gases will form compounds with other elements. *Krypton* is derived from the Greek word for *hidden*, since it was a game of hide-and-seek to find it.

Lanthanum La (At. Wt. 138.9, At. No. 57)

	4925.2 Fe (1000)	(200) La 4703.2	
(500) La 4921.7 Ti (100)		(200) La 4699.6	
	4920.5 Fe (500)	(200) La 4692.5	
(500) La 4920.9			
	4918.9 Fe (300)		4337.0 Fe (400)
		(800) La 4333.7	
(100) La 4748.7			4332.9 Ba (20)
(300) La 4743.0			4325.7 Fe (1000)
(150) La 4740.2			
(400) La 4728.4		P(300) La 6249.9	
	4722.1 Zn (400)	P(1000) La 3949.1	

An easily recognized lanthanum group 4728-4748 lies within the brilliant zinc triplet. The strong doublet La-4920-4921 stands out clearly in the visible spectrum. The triplet 4692-4699-4703 with lines of equal intensity and characteristic spacing is also easily spotted. Although its intensity is much lower, the orange line La-6249 is more persistent than La-3949 in the ultraviolet region.

Lanthanum and cerium, both discovered by Moslander in 1839, were the first of the rare earths to be known. Later he discovered terbium and erbium. Lanthanum, one of the more plentiful and better known rare earths, has a metallic luster, and, like iron is ductile and malleable; like magnesium it burns in air. In the minerals cerite, gadolinite and allanite it occurs as a silicate; in lanthanite, as a carbonate. Lanthanum was first found in cerite from Sweden, but it has since been found in many places, including the leaves of hickory trees growing over pegmatite veins. It combines with oxygen to form a sesquioxide and also with nitrogen, chlorine, and sulfur to form colorless astringent salts. Unlike most of the rare earths, it gives no absorption spectrum.

Characteristic Lines of the Elements

Samples of lanthanum salts are sometimes available for a few cents. The price of the metal is $3.50 per gram.

Lead Pb (At. Wt. 207.21, At. No. 82)

H(40)	Pb 6001.8		4062.6	Cu	(500)
(10)	Pb 5201.4	(20)	Pb 4062.1		
(20)	Pb 5005.4		4058.9	Cb	(1000)
		PR(2000)	Pb 4057.8		
PR(500)	Pb 2833.0		4055.2	Ag	(800)
PR(1000)	Pb 2169.9		4047.2	K	(400)

Lead is most reliably identified by the remarkable violet couplet 4057-4062 with one line a hundred times the intensity of the other and with strong copper, silver and potassium lines as spotters. Lead with three lines having wave-lengths close to even hundreds of angstroms, as shown above (6001.8, 5201.4, 5005.4), is a good element to use for plotting and testing scales. Pb-6001 is also valuable for identification, since it is much more prominent in the 110-volt arc than its table intensity would indicate.

The weight of lead was observed in Biblical times, about 580 B.C., for it is stated, Exodus 15:10, that the Egyptians "sank as lead in the mighty waters." In Ezekiel, 22:20, something of the ancient method of smelting is indicated: "they gather silver and brass and iron and lead and tin into the midst of the furnace to blow the fire upon it to melt it." The process was not so different from the modern process of roasting.

Lead is malleable, has a low melting point, and is soft. It is a poor conductor of heat and of electricity, and is very resistant to acids. The chief ore is galena, PbS, which contains over 86 per cent lead. Other ores of less importance are raspite with arsenic, crocoite with chromium, wulfenite with molybdenum and cerussite a carbonate. As indicated in the Bible, silver, copper and zinc ("brass") are frequently found in lead ores. Antimony, arsenic and gold sometimes are also present. Lead has been found in commercial deposits in Missouri, Iowa, Illinois, Colorado and Utah.

So great is its demand for storage batteries, ethyl gasoline, plumbing, type metal, etc., that it is not likely to remain long at the pre-war price of $0.05 or $0.06 per pound.

LITHIUM LI (AT. WT. 6.94, AT. NO. 3)

6717.6 Ca	(500) Li 4971.9 Co (150)
PR(3000) Li 6707.8	
6693.8 Ba (600)	4607.3 Sr (1000)R
	P(800) Li 4602.8 Fe (300)
6112.7 Ba (200)H	
PR(2000) Li 6103.6	PR(1000) Li 3232.6
6102.7 Ca (90)	

The strongest and most persistent lithium line 6707 is almost as omnipresent as the sodium doublet. It will ordinarily be the only red line in the vicinity in rock analysis, except for the frequently occurring Ca-6717 on one side and Ba-6693 on the other. With 110 volts the line shows reversal at about 5 per cent, and it is so sensitive that Bunsen declared he could detect the one sixty-thousandth part of a milligram of lithium in his flame. Li-6103 is another strong line very close to Ca-6102. Li-4602 is not far from the very sensitive strontium line 4607. Like those of lead, the strong lines of lithium lie near the even hundred angstrom marks and so are very valuable in the calibration of spectroscopic scales.

Lithium was discovered by Arfvedson in 1817. It is found in nearly all rocks, in tobacco ash, in many plants, in meteorites and in the sun. In most of its occurrences it is only as a trace, but in spodumene, lepidolite, and amblygonite it is found in quite appreciable amounts. Oxides, carbonates, chromates, nitrates, the citrate, the chloride and many other compounds are known.

Lithium, the lightest of the metals, has about half the density of water, it is about one-third as heavy as magnesium, and only one-fifth as heavy as aluminum. Its price is about $20 per pound, that of the carbonate $2 per pound.

Characteristic Lines of the Elements

LUTECIUM LU (AT. WT. 174.99, AT. NO. 71)

6016.6 Fe (100)	5476.9 Ni (400)
6013.5 Fe (100)	(500) Lu 5476.6
(400) Lu 6004.5 Eu (300)	5474.9 Fe (100)
6001.8 Pb (40)	5473.9 Fe (100)

P(300) Lu 4518.5
P(100) Lu 2911.3

Lutecium has fewer spectral lines than any of the other rare earths. Its spectrum contains about 500 lines—and, most of those are very weak. However, there are a few fairly strong, well placed lines for visual identification. Lu-6004, one of these, has a close iron doublet for spotting and also a lead spotter. Lu-5476, a little stronger, also has an iron doublet close at hand and a nickel line so close as to require checking. Lu-4518 is persistent, but without convenient spotters and with titanium interference. The ultraviolet lines are not so strong as the visible ones.

Lutecium, discovered by Urbain in 1907, had up to that time been so closely associated with ytterbium as to be considered a part of it. A year later Welsbach confirmed the separation. The white sesquioxide, the chlorate, and the sulfate have all been prepared, but the pure metal has not been isolated.

At first the rare earths, so similar in characteristics, were separated only by fractional precipitations, or by tedious chains of hundreds or even thousands of crystallizations before the substance was sufficiently pure; more recently an electrolytic method which greatly reduced the time and labor of the separation was developed.

MAGNESIUM MG (AT. WT. 24.32, AT. NO. 12)

5192.3 Fe (400)	(60) Mg 5528.4
5191.4 Fe (400)	
5188.8 Ca (50)	(8) Mg 4703.2
H(500) Mg 5183.6	
5181.9 Zn (200)	P(300) Mg 3838.2
5173.7 Ti (125)	P(100) Mg 3829.3

H(200) Mg 5172.6 P(250) Mg 3823.3
      5171.5 Fe (300)
      5167.4 Fe (700)   PR(300) Mg 2852.1
H(100) Mg 5167.3
      5166.2 Fe (125)

Magnesium is most easily recognized by its triplets, of which it has one in the visible and another in the ultraviolet region; the latter is a miniature replica of the former. Two lines of the visible group Mg-5167-5172-5183 are spaced very much like the lines of the sodium doublet, with the third twice as far away. When only traces of magnesium are present, the intermingling iron lines may cause some difficulty, but usually the magnesium lines, characteristically a little hazy, are easily distinguishable. In dolomite, Ca-5188 joins with the triplet to make a symmetrical quartet indicative of that mineral. In the 110-volt arc Mg-4703, intensity only (8), often comes out brilliantly. The most sensitive line is Mg-2852, in the ultraviolet region. Magnesium, constituting over 2 per cent of the earth's crust, occurs in a great many rocks. Not more than half a dozen spectra are more frequently seen.

Magnesium was another of the elements first isolated by Davy in 1808. It is a brilliant, silvery metal, lighter in weight than aluminum or any other of those in common use. Before World War II about the only use of the pure metal was for flashlight powders and fireworks, but the need for a light metal in airplane construction led to the erection of great magnesium plants and the industry promises to remain a permanent one. The metal may be produced from either magnesite or brucite but dolomite has so far been most popular, because of the very large deposits available. Since it constitutes appreciable proportions of serpentine and sea water, its sources are inexhaustible.

The 1944 price was $0.20 per pound.

Manganese Mn (At. Wt. 54.93, At. No. 25)

(400) Mn 4823.5   (80) Mn 6021.7 Fe (300)
      4821.0 Fe (200)   (80) Mn 6016.6 Eu (1000)
      4810.5 Zn (400)   (100) Mn 6013.4 Fe (100)
(400) Mn 4783.4

Characteristic Lines of the Elements 135

(80)	Mn	4766.4		P(250)	Mn 4034.4
(60)	Mn	4765.8		P(400)	Mn 4033.0
(100)	Mn	4762.3			4032.9 Ga (1000)
		4759.2	Ti (100)	P(500)	Mn 4030.7
		4758.1	Ti (100)		
		4756.1	Cr (300)	PR(200)	Mn 2593.7
(400)	Mn	4754.0		PR(300)	Mn 2576.1

The strongest and most persistent manganese group is Mn-4030-4033-4034, but a group which is placed more conveniently for visual observation and is also clearer of interference is Mn-4754-4783. Within this group is the titanium doublet Ti-4758-4759, which serves as a spotter. The weaker red triplet Mn-6013-6016-6021, not very far from the sodium doublet, is also easily recognized if any considerable amount of the element is present in the sample. Note also the persistent triplet in the ultraviolet region.

Manganese, one of the first of the new metals to be brought to light in modern times, was isolated by Gahn in 1774. Black manganese ores previously had been called "lapis magnesius" and were supposed to contain iron. Scheele claimed that some other metal was present and Gahn confirmed Scheele's work by isolating the element.

Manganese is a hard, brittle metal with a melting point slightly below that of iron. It is not used in the metallic state, but it forms iron alloys of great tensile strength and elasticity. It may be used in almost any proportion, from 10 to 80 per cent, to form alloys suitable for various uses such as frames, springs, or automotive bearings. Carbon is always present in manganese steels.

The chief ores are pyrolusite, psilomelane, manganite and rhodochrosite (pink). Considerable bodies of ore are found in the United States, but the great deposits are in Russia and Brazil. The price of 48 per cent ore was $48 per ton in 1944.

MASURIUM MA (AT. WT. 99, AT. NO. 43)

K(a) Ma $\begin{cases} .7078 \text{ Mo K(a) X-ray}^* \\ .6417 \text{ Ru K(a) X-ray} \\ .6120 \text{ Rh K(a) X-ray} \end{cases}$ K(b) Ma $\begin{cases} .6309 \text{ Mo K(b) X-ray} \\ .5713 \text{ Ru K(b) X-ray} \\ .5444 \text{ Rh K(b) X-ray} \end{cases}$

* See footnote under ILLINIUM.

The only evidence of the existence of masurium is its X-ray spectrum, first observed in 1925 by Noddack, Tacke and Berg in columbite. It occurs also in sperrylite, and the rare earth minerals. Its proportion in the earth is estimated to be but one part in a trillion.

X-ray spectral lines are of much shorter wave-lengths than those of light, as indicated above. In the K-series the wavelengths of the spectral lines of the elements decrease as the atomic numbers increase, and the lines of each element fall into their places in exact order. In the present instance, the lines of masurium, No. 43, fall between those of molybdenum, No. 42 and those of ruthenium No. 44. When these lines were discovered it was known that the long-missing element ekamanganese had been found—even though it was present in too minute a quantity to be otherwise detected.

MERCURY HG (AT. WT. 200.6, AT. No. 81)

	5793.5	C	(Band)	P(3000) Hg	4358.3	Fe	(70)
(600) Hg	5790.6				4352.7	Fe	(300)
	5785.9	Ti	(100)H		4351.7	Cr	(300)
	5782.1	Cu	(1000)	(200) Hg	4347.4	Cr	(200)
	5777.6	Ba	(500)R		4344.5	Cr	(400)
	5774.0	Ti	(70)H	P(200) Hg	4046.5		
(600) Hg	5769.5						
	5766.3	Ti	(70)H	PR(2000) Hg	2536.5		

Mercury lines appear very quickly and also vanish quickly, due to the rapid evaporation of the metal. They should consequently be looked for early in the survey. The visible line Hg-4358 is persistent and brilliant. The doublet Hg-5769-5790 is also very brilliant in the 110-volt arc. Cu-5782 lies about midway between its components. Other persistent lines lie in the ultraviolet region.

Mercury was known to the ancients, who obtained it by roasting cinnabar with charcoal. It was also of great interest to the alchemists of the Middle Ages who distilled it and believed it to be a sort of essence of other metals.

Characteristic Lines of the Elements

Mercury is liquid at ordinary temperatures, but when cooled, as by pouring it into liquid air, it becomes hard like other metals and, a hammer molded from it, may be used to drive nails. Its chief ore is cinnabar, the sulfide, from which it is separated by roasting and distillation. The chief deposits are in Spain, Austria, and California.

During wars, mercury is always in great demand for fulminating caps to set off charges of high explosives. In World War II, the chief United States supply was ex-president Hoover's New Idria mine of San Benito County, California, although some was also derived from New Almaden, Guadalupe, and Oregon.

In times of peace, mercury is used in thermometers, in the concentration of gold, in mercury-vapor lamps, in mirror backs, in medicine and in scientific apparatus. A recent price was $175 per 76 pound flask, but it has dropped since the war.

Molybdenum Mo (At. Wt. 95.95, At. No. 42)

	5572.8	Fe	(300)		(300) Mo 6030.6	
(200)	Mo 5570.4				6021.8	Fe (300)
	5569.6	Fe	(300)			
	5535.5	Ba	(1000)	PR(1000) Mo 3864.1		
(200)	Mo 5533.0			PR(1000) Mo 3798.2		
(200)	Mo 5506.4	Fe	(150)			

The persistent and most brilliant molybdenum lines are in the ultraviolet region, but there are also strong visible lines of which Mo-6030 is the most easily spotted, as it is ordinarily free from interference. The three yellow lines Mo-5506-5533-5570 usually stand out prominently in ore of any value, but if the ore is too low-grade, iron lines often confuse the picture. For mere traces, Mo-6030 is more reliable than the triplet.

Molybdenum ores were long mistaken for those of lead, but in 1778 Scheele showed that they contained a new substance. Four years later Hjelm confirmed his findings by isolating the new metal. Since the ores of lead, graphite, and molybdenum may all be black and have a greasy feel, they often puzzle modern prospectors as much as they did the ancients.

Molybdenite MoS_2, the chief ore, is found in granite and various crystalline rocks. It is a soft, gray-black, laminated mineral, containing about 40 per cent sulfur, and often occurs in schist. Wulfenite, a lead molybdate, is another important ore; it is often found with vanadinite. Molybdenum is found in New York, Colorado, and Canada. Deposits also occur near Barstow, California. The metal is soft and malleable, but it may be tempered. It has a great variety of uses, e.g., X-ray electrodes, rectifiers and resistors, but its chief value is in giving hardness to steels, especially those used for machine tools, shafts, and drills.

High-grade (90 per cent) molybdenite brings $0.45 per pound.

NEODYMIUM Nd (AT. WT. 144.27, AT. NO. 60)

	6400.0 Fe (200)			5895.9 Na (5000)	
	6393.6 Fe (100)		(20) Nd 5891.5		
(100) Nd	6385.1			5889.9 Na (9000)	
	6362.3 Zn (1000) H				
				4305.4 Fe (100)	Ti (300)
	5624.5 Fe (150)	P(100) Nd	4303.5 Pr (100)		
(200) Nd	5620.5			4302.5 Ca (50)	Fe (50)
(10) Nd	5619.0 Ba (10)			4301.0 Ti (150)	
	5615.6 Fe (400)			4300.5 Ti (125)	
(150) Nd	5594.4 Ca (35)			4299.2 Fe (500)	

Neodymium has over 2500 spectral lines and none of them is very bright. They are often hard to distinguish from the stronger lanthanum lines which occupy the same spectral regions. In the absence of praseodymium, it may be spotted by its most persistent line, Nd-4303. To avoid error, the identification should be checked by the stronger yellow line, Nd-5620; this is in a less crowded field, which also contains Nd-5594 as further confirmation. There is, besides, a weak neodymium line between the two components of the sodium doublet; this may be seen if the sodium lines are not too brilliant. It frequently happens, however, that the lines of the doublet are so overpowering as to blot out not only the many lines between them but those for some distance on either side as well.

Neodymium, a rare earth, was one of two discovered by Welsbach in 1885; he added a third, lutecium in 1907. In 1843 Mosander, the discoverer of four of the rare earths, reported the discovery of *didymium*; which was unchallenged as an element for 42 years. Then Welsbach found that it really consisted of two elements, which he called neodymium and praseodymium, almost inextricably bound together. He managed to separate them by repeated fractionation of the nitrates. It was the fourth rare earth to be extracted from cerite, and still others were to follow.

Neodymium ammonium nitrate is priced at $0.25 per gram, the metal at $3.50 per gram.

NEON NE (AT. WT. 20.183, AT. NO. 10)

P(2000) Ne 6402.2 Electron Tube
P(2000) Ne 5852.4 Electron Tube
P(2000) Ne 5400.5 Electron Tube

Neon has about 2000 spectral lines, the average number for an inert gas; they are also of average brilliance. The preponderance of red and yellow light waves gives the neon tube the dazzling color used to advantage in advertising.

Neon was one of three inert gases discovered by the renowned Scotch chemist, Ramsay, in 1898. He had previously discovered argon and helium and was later awarded the Nobel prize for his work with the inert gases. After first liquefying air he secured neon by fractional distillation.

Neon, much less abundant than argon, constitutes but one part in 66,000 of air; it forms no compounds and liquefies at $-205°C$ under a compression of 29 atmospheres. In the open air it soon boils away and mixes once more with the other gases of the atmosphere. The neon tube is a more efficient source of light than tubes filled with other gases, but because of the irritating red color it is not adapted for domestic lighting. If the gas is not pure, brilliance is greatly diminished; complete separation is made possible by the absorptive power of carbon.

For commercial signs a transformer is used to step up the

110-volt, 60-cycle, alternating current to 6000 volts. Simple neon tubes a foot in length are sometimes used to test automobile spark plugs. Each time the plug is energized the tube glows and shows the efficiency of the spark coil. Such tubes will also glow from static electricity when rubbed in the dark.

The price of neon at atmospheric pressure is $10.00 per liter. When it is rarefied, a small amount of gas will suffice for many tubes.

NICKEL NI (AT. WT. 58.69, AT. NO. 28)

(500)	Ni 6191.1			(500)	Ni 4984.1	
(400)	Ni 6176.8				4983.8 Fe	(200)
(300)	Ni 6175.4				4982.8 Na	(200)
	6169.5 Ca	(65)			4982.5 Fe	(200)
					4981.7 Ti	(300)
H(400)	Ni 5476.9			(500)	Ni 4980.1	
	5474.9 Fe	(100)				
	5473.9 Fe	(100)			4722.1 Zn	(400)
	5471.5 Ag	(500)		(200)	Ni 4715.7	
	5465.4 Ag	(1000)		(1000)	Ni 4714.4	
(300)	Ni 5084.0				4404.7 Fe	(1000)
(100)	Ni 5082.3			(1000)	Ni 4401.5 Fe	(60)
(150)	Ni 5081.1					
(200)	Ni 5080.5			PR(1000)	Ni 3492.9	
				PR(1000)	Ni 3414.7	

Nickel has over 1000 lines of about the same intensity as those of iron. Ni-5476 (yellow) is easily spotted just two angstroms from a close iron doublet. A bit of silver chloride on the sample will also bring out the check lines Ag-5465-71. A more reliable check perhaps, is the doublet Ni-4714-4715 (blue) which can be spotted near Zn-4722, produced by burning a little zinc oxide in the arc. Another of the very strong lines is Ni-4401 (blue) near Fe-4404. Also note the strong green doublet Ni-4980-4984, and the crowded quadruplet 5080-5081-5082-5084 with lines closer together toward the yellow.

Nickel was the second of the many new metals to become known in modern times. It was discovered in 1751 by Cronstedt

16 years after the isolation of cobalt. Although one of the better-known metals, nickel is very selectively distributed and its spectral lines are seldom encountered. The most common ores are niccolite, garnerite, and millerite; the most valuable mines are in Canada, Caledonia, and Spain.

The alloy, nichrome, is used in electric heating elements; the metal is used for electroplating, for magnets in stainless steels and for cooking utensils.

The prewar price of nickel was $0.35 per pound.

NITROGEN N (AT. WT. 14.008, AT. No. 7)

P(500) N 5679.5 Discharge Tube
P(300) N 5666.6 Discharge Tube
P(1000) N 4109.9 Discharge Tube

The persistent lines of nitrogen are all visible. It has 832 ordinary spectral lines, only a few of them very bright.

Nitrogen, which constitutes about four-fifths of air and of man's every breath, was discovered by Rutherford in 1772. Free nitrogen is found in air, in volcanic gases, and in the atmosphere of the sun. In combination, it is found in many natural ores and minerals.

It is an essential constituent of the tissues of all living organisms, both plant and animal. One of the chief problems of agriculture is to replenish the supply of nitrogen in the soil. This may be accomplished by crop rotation, introducing every third year or so a clover or a legume whose roots nourish nitrogen-accumulating organisms. Plants can use only soluble compounds of nitrogen, and not the free nitrogen of the air. In recent years nitrogen has been extracted directly from the air in the hydroelectric fixation plants of Norway and of Niagara Falls.

Nitrogen has been both liquefied and solidified. Although it is sluggish chemically, it forms many compounds, among which nitrous oxide, *laughing gas,* discovered by Priestly, is used as an anesthetic. Other important compounds are nitric acid, so familiar in chemical laboratories; nitro-benzene, derived from coal-tar, and used in the manufacture of dyes; nitro-cellulose, used in explosives and in plastics; and nitroglycerin, another high explosive.

Both sodium and potassium nitrate sell at about $0.30 per pound.

OSMIUM Os (AT. WT. 190.2, AT. No. 76)

	4799.9 Cd (300)	4554.0 Ba (1000)
(300) Os 4793.9		4554.4 Ti (150)
	4792.8 Co (600) Au(200)	(150) Os 4551.2
	4792.5 Cr (200)	(150) Os 4550.4
	4791.2 Fe (200)	4549.6 Ti (100)
		(100) Os 4548.7 Ti (125)
	4425.4 Ca (100)	4547.8 Fe (200)
	4422.5 Fe (300)	
P(400) Os	4420.4 Sm (200)	PR(500) Os 2909.0
	4415.1 Fe (600)	

The strongest and most persistent visible osmium line is Os-4420, half way between the blue iron and calcium groups so often seen in common rocks and minerals and so easily recognized because of their symmetry. Probably the most distinctive osmium group is Os-4548-4550-4551, surrounded by iron and titanium lines. There is another strong line near Cd-4799. The most persistent line is in the ultraviolet region.

Osmium, along with iridium, was discovered by Tennant in 1803 in the residue after extracting all platinum from some ore. The members of the platinum group are found almost as closely associated in nature as the rare earths.

Osmium is nearly 22.5 times as heavy as water and is the heaviest of the group. It is also heavier than lead (specific gravity 11.35) or gold (specific gravity 19.32). In fact, it is heavier than any other of the elements, or of their known alloys.

The name osmium is derived from the Greek word meaning odor and the element was so called from the pungent vapor accompanying its oxidation. Its uses are the same as those of iridium.

The price of osmium is $75.00 per ounce.

Characteristic Lines of the Elements 143

OXYGEN O (AT. WT. 16, AT. NO. 8)

P(100) O 7775.4 Discharge Tube
P(300) O 7774.1 Discharge Tube
P(1000) O 7771.9 Discharge Tube

The persistent lines of oxygen are visible but close to the border of the infrared region. They may be seen at the red end of the solar spectrum, where they usually appear as a black band. There are about 440 oxygen lines above intensity (1).

Oxygen was first isolated by Priestly in 1774, two years after the discovery of nitrogen by Rutherford. A year later, Scheele independently made the discovery in Sweden, and later still Lavoisier, in Paris, made the first completely satisfactory analysis of air by showing it to be a mixture of the two gases, oxygen and nitrogen. (The existence of the inert gases was not even suspected at that time.)

Oxygen constitutes about half of the whole earth including rock, air, and water.

It may be liquefied at −118° and under 50 atmospheres pressure. It may be obtained by merely placing two electrodes in water and collecting the gas which bubbles from the anode, or by heating a mixture of potassium chlorate and manganese dioxide in a test tube and collecting the gas driven off. Oxygen is very active and combines with nearly all the other elements except the inert gases. It is used in welding, medicine, high-altitude flying, etc.

The price of oxygen is $0.50 per cubic foot.

PALLADIUM Pd (AT. WT. 106.7, AT. NO. 46)

	5675.7 Na	(150)		4213.6 Fe	(100)
(100) Pd	5670.0 Na	(100)H	(500) Pd	4212.9	
				4210.3 Fe	(300)
	5543.1 Fe	(25)	(500) Pd	4087.0	
	5165.2 C	(Band)	PR(2000)	Pd 3421.2	
	5164.6 Fe	(70)H	PR(2000)	Pd 3634.6	
(300) Pd	5163.8				
	5162.2 Fe	(300)			

Palladium is best identified in the ultraviolet region, where its most brilliant spectral lines are to be found. The strongest visible line, Pd-4212, may be spotted by means of Fe-4210. Pd-5670, near the greenish yellow sodium doublet Na-5670-5675, is easily located. Pd-5163 may be spotted by the carbon band nearby.

Palladium, as well as rhodium, was discovered by Wollaston in 1803. He named the new element after the planetoid discovered the previous year. It is the third member of the platinum group consisting of the six elements listed below. They are arranged according to atomic numbers.

TABLE 16

THE PLATINUM METALS

Element	Atomic Number	Discovered by	Discovered in
Ruthenium	44	Claus	1845
Rhodium	45	Wollaston	1803
Palladium	46	Wollaston	1803
Osmium	76	Tennant	1804
Iridium	77	Tennant	1804
Platinum	78	Wood	16th Century

Four of the six were discovered within two years. Platinum had already been known for more than 100 years. The members of the family are all found in the same ores and sands, and most of them were discovered in platinum residues.

The metals, singly or alloyed, usually occur native in alluvial sands—sometimes in river beds, the natural sluices of nature, sometimes on ocean beaches. The California platinum sands are black and usually richer in gold than in the platinum metals.

Palladium, in the form of spongy palladium and palladium black, will absorb tremendous quantities of hydrogen and is used as a hydrogenation catalyst.

The price of palladium is about $24.00 per ounce.

Phosphorus P (At. Wt. 30.98, At. No. 15)

Arc	(300)	P	4222.1		
Arc	P(60)	P	2554.9	P(20)	Electron Tube
Arc	P(80)	P	2553.2	P(20)	Electron Tube
Arc	P(100)	P	2535.6	P(30)	Electron Tube
Arc	P(50)	P	2534.0	P(20)	Electron Tube

The spectral lines of phosphorus are weak and unsatisfactory, in both the visible and ultraviolet regions. Only under special conditions can the element be determined spectrographically. P-4222, listed as visible in some tables, does not appear in the 110-volt arc.

Phosphorus, even in small quantities, may be detected chemically by dissolving the specimen in nitric acid and adding ammonium molybdate to the solution. If phosphorus is present, a fine yellow precipitate comes down. Insoluble ores must be fused with sodium carbonate; the product is leached with water, and the solution is filtered. The filtrate carries the phosphorus, along with silica and other elements. Ammonium molybdate is then added to the nitric acid solution of the orthophosphate to precipitate the yellow phosphomolybdate.

Phosphorus was first discovered in urine by Brand in 1669. Two years later Scheele prepared it from bones. There are several allotropic forms; the most common are the red and white modifications. The white form has a yellowish tint and is a violent poison; a fraction of a gram constitutes a fatal dose. It is used in roach and rat poisons. It is phosphorescent in air and ignites spontaneously; it is usually stored under water for this reason. Red phosphorus, obtained by heating the white form in vacuum, is comparatively non-toxic and non-reactive, and need not be kept under water. It is used in match heads.

The chief ores are the phosphates and phosphorites, apatite being used as a soil fertilizer. Phosphorus is an essential ingredient of living tissues and is found particularly in bone, muscle and nerve tissue. Eggs, beans, nuts and whole wheat are rich food sources of phosphorus.

Red phosphorus is priced at about $1.00 per pound.

PLATINUM PT (AT. WT. 195.2, AT. NO. 78)

	6767.7 Ni (300)			5302.8 Ba (20)	
(100) Pt 6760.0				5302.3 Fe (300)	
			(150) Pt 5301.0 Co (700)		
(50) Pt 5478.4					
	5477.7 Ti (70)		(800) Pt 4442.5		
	5476.5 Fe (80)			4442.3 Fe (400)	
(60) Pt 5475.7					
	5474.9 Fe (100)			4418.7 Co (1000)	
	5473.9 Fe (100)		(600) Pt 4418.6		
				4418.5 Fe (400)	
PR(2000) Pt 3064.7					
PR(2000) Pt 2659.4					

Platinum has 1000 spectral lines—fewer than the other metals of its group. The strongest lines are in the ultraviolet region. Pt-4442 has a strong iron line unfortunately close; Pt-4418 has a still closer iron line. In the presence of iron, therefore, it is necessary to find a weaker line such as Pt-5301, more than an angstrom from its iron companion, and thus resolvable by many spectroscopes. The platinum doublet Pt-5475-5478 is in a compact group of iron and titanium lines, which may be analyzed as follows: a single line means titanium; a doublet, platinum; a triplet, platinum and titanium or iron alone,* depending upon the spacing; a quadruplet, iron and titanium; a quintuplet, iron and platinum; a sextet, all three elements present.

Platinum was first introduced in Europe by Don Antonio de Ulloa, who brought some of it back from Colombia in 1735. It was described by Scheffer in 1751 as "a perfect metal as stable as gold or silver."

Russia is the chief source of the metal, and nuggets weighing as much as four pounds, have been found there. Colombia still

---

* In the group marked at the top of the page:
Two platinum lines and one Ti would make a triplet group.
Three Fe lines would make a triplet group.
There are no other combinations of 1, 2, and 3 that would make a triplet.
If the two lines toward the violet are close together the triplet means iron.
If the two top lines are close together—Pt and Ti.

supplies considerable amounts, and California some. It is the most ductile of metals, and is highly resistant to corrosion. Its chief use is for scientific equipment. Its recent price is about the same as that of gold.

POLONIUM Po (AT. WT. 210, AT. No. 84)

(20) Po 2558.1 Discharge Tube

Polonium has two or three known spectral lines; but since, like other radioactive elements, it may better be detected by electroscopes or Geiger-Müller counters, the spectrum is not important.

Polonium was discovered in 1898 by Madam Curie, when she found some pitchblende that was unusually active. It resembled bismuth so closely that for a time others thought it was only an alloy of that metal. It was later observed that when the alpha particles from polonium hit beryllium or lithium a new type radiation was produced. In 1932 Irene Curie and Joliot observed that if the rays struck paraffin or other carbon compounds, high energy particles were ejected. Finally Chadwick in England showed that these particles were of the mass of protons but without an electric charge, and so called them *neutrons*. These facts later became important in the development of atomic energy.

Polonium subnitrate is a whitish powder hundreds of times more radioactive than uranium nitrate; its emanations induce fluorescence.

POTASSIUM K (AT. WT. 39.096, AT. No. 19)

P(5000)	K 7698.9		4055.2 Ag (800)
P(9000)	K 7664.9	P(400)	K 4047.2
			4046.5 Hg (200)
(500)	K 6938.9		4045.8 Fe (400)
(300)	K 6911.3	P(800)	K 4044.1
(50)	K 5832.0		
(30)	K 5812.5		

There are less than 300 lines of higher intensity than (1) in the spectrum of potassium, but the deep red lines of the doublet K-7664-7698 equal in intensity those of the sodium doublet Na-5889-5895, and are among the strongest lines encountered in spectroscopy. Like those of sodium, the potassium lines are arranged in pairs with the components drawing closer together toward the ultraviolet region. Ordinarily, the couplets K-7664-7698 and 4044-4047 are the most convenient lines for observation. The persistent lines are all visible.

Potassium was first isolated in 1807 by Sir Humphry Davy, discoverer or co-discoverer of a half dozen other elements. He was a brilliant chemist and lecturer, president of the Royal Society of London, knighted for his scientific achievements. He used a voltaic pile of 200 plates and the electrolytic method in making most of his separations.

The metal reacts violently with water and oxidizes quickly in air. Potassium is found in all land plants and in their ashes. Chemically it acts almost exactly like sodium. It is used in making glass, gunpowder, fertilizers, drugs and in numerous chemical compounds. The carbonate may be recovered from wood ashes by leaching; its nitrate occurs as the natural mineral, saltpeter.

The price of the pure metal is about $2.00 per ounce, that of the bromide $0.33 per pound.

PRASEODYMIUM PR (AT. WT. 140.92, AT. NO. 59)

(125) Pr 4783.3
(100) Pr 4744.9
(80) Pr 4744.1
      4743.6 Gd (300)
      4743.0 La (300)
(125) Pr 4736.6 Fe (125)
(100) Pr 4734.1

      4480.3 Cu (200)
(125) Pr 4477.2
      4476.0 Fe (500)

      4433.2 Fe (150)
(80) Pr 4432.3
(60) Pr 4431.8 Ba (60)
      4430.6 Fe (200)
(200) Pr 4429.2
      4427.3 Fe (500)

      4191.4 Fe (200)
P(100) Pr 4189.5
      4187.8 Fe (450)
      4181.7 Fe (200)
P(200) Pr 4179.4

Praseodymium is probably most easily recognized by the blue doublet Pr-4734-4736 with the triplet Pr-4744-4744-4783 only eight angstroms away. The most persistent lines are Pr-4179-4189 in the violet region. One of its strongest lines, Pr-4429 has a weaker doublet nearby. Pr-4477, flanked on one side by copper (Cu-4480) and on the other by iron (Fe-4476), is easily spotted. Altogether it has over 2,700 lines, none very strong.

Praseodymium was discovered in 1885 by Welsbach, also the discoverer of neodymium. Both are rare earths; their mixture was originally believed to be the element "didymium." Both were obtained from cerite, a complex silicate in which lanthanum and other rare earths are also found. Its salts are green, it has a characteristic absorption spectrum and it belongs to the cerium group. Cerium and lanthanum were discovered first; then, by further break-down the neodymium, praseodymium and illinium chain was developed, and also the samarium, gadolinium and europium chain. Few of the fifteen rare earths, atomic numbers 57 to 71, are procurable in sufficient quantity to have much practical importance. A list of the rare earths is in Table 14.

Praseodymium oxide is listed at $1.00 per gram.

RADIUM RA (AT. WT. 226.05, AT. NO. 88)

P(800) Ra 4825.9 Electron Tube
P(800) Ra 4682.2 Electron Tube
P(2000) Ra 3814.4 Electron Tube

The radium spectrum is usually studied in a discharge tube, but its arc spectrum is very similar to that of the discharge tube. All the lines given above, for instance, appear also as strong arc lines. Relatively few radium lines (140) are catalogued.

Bacquerel in 1896 discovered the radioactivity of uranium. Jacques and Marie Curie, greatly interested, decided to investigate the matter and as a result of concentrations from tons of pitchblende discovered radium, forty times as radioactive as uranium. Demarcay examined its spectrum and agreed that it was a new element. A bit of a radium salt placed a fraction of a centimeter above a fluorescent screen and viewed through a lens of one-inch focus in a darkened tube, will show ceaseless scintillations as the alpha particles strike the screen. Such an instrument is called a spinthariscope and the display is very impressive.

The discovery of radioactivity led directly to the modern atomic theories of Rutherford and Bohr, to the invention of the mass spectrograph, the determination of isotopes, and a whole new conception of matter and energy. It suggested Einstein's theory of the possibility of converting a small amount of matter into an immense amount of energy. Eventually it led to the harnessing of atomic power.

The radiations of radium eventually proved to consist mainly of alpha particles and also some of the beta and gamma types. The alpha particle has been shown to be a helium nucleus consisting of two protons and two neutrons, all having about the same mass. The beta radiations consist of electrons, the gamma radiations are electromagnetic in character, similar to X-rays. The average life of radium is computed as 2440 years; at the end of this time most of the radium will have been transmuted into lead and helium.

Radium is used in the treatment of tumors and cancers and in selfluminous paints.

The price is $20.00 per milligram.

Radon Rn (At. Wt. 222, At. No. 86)

P(600) Rn 7450.0 Geissler   (300) Rn 5084.5 Geissler
P(400) Rn 7055.4 Geissler   (500) Rn 4680.8 Geissler

The spectrum of radon has 420 lines, with the persistent ones close to the infrared region. Because of its radioactivity it is best detected with a Geiger-Müller counter.

Radon was discovered by Dorn in 1900, and isolated by Ramsay and Gray in 1908. It is the last gas in the inert group, and the heaviest of all gases. It is a decomposition product of radium and has an average length of life of less than a week. It is col-

Characteristic Lines of the Elements

lected in tubes and used like radium in the treatment of cancer. The price is $2.00 per millicurie.

RHENIUM RE (AT. WT. 186.31, AT. NO. 75)

```
            5283.6  Fe  (400)                  5782.1  Cu  (1000)
            5281.7  Fe  (300)                  5777.6  Ba  (500)
(100)  Re   5278.2                     H(300) Re 5776.8
(500)  Re   5275.5
H(200) Re   5270.9  Fe  (400)          (300)  Re 5377.0
            5269.5  Fe  (800)                  5371.4  Fe  (700)
            5266.5  Fe  (500)
            5263.3  Fe  (300)          HP(1000) Re 4889.1
                                       HP(1000) Re 3460.4
```

The strongest rhenium line Re-4889 is in a rather crowded field, but is well placed for observation. The next strongest line is in the ultraviolet region. The rhenium doublet Re-5275-5278 fortunately occurs in a large gap in a strong iron group. Rhenium was discovered by Noddack, Tack and Berg in 1925. It occurs in minute quantities in pegmatites with columbian, tantalum and tungsten. Originally considered very rare, it was later produced in considerable quantity. Meggars found over 2000 lines in its spectrum. It is a member of the manganese group and is estimated to constitute only one one-billionth of the lithosphere. The price has varied from $3.00 to $8.50 per gram.

RHODIUM RH (AT. WT. 102.91, AT. NO. 45)

```
           5615.6  Fe  (400)                  4378.2  Cu  (200)
           5601.2  Ca  (15)                   4375.9  Fe  (500)
(300) Rh   5599.4                             4374.9  Mn  (150)
           5598.4  Ca  (35)          H(1000) Rh 4374.8
           5586.7  Fe  (400)                  4369.7  Fe  (200)

                                              4289.7  Cr  (3000)
           4530.8  Cu  (200)                  4289.0  Ti  (125)
(600) Rh   4528.7                    (400)  Rh 4288.7
           4528.6  Fe  (600)                  4288.1  Fe  (50)
           4526.9  Ca  (100)                  4287.4  Ti  (100)
                                     P(1000) Rh 3434.8
```

The strongest visible rhodium line Rh-4374 is very close to a manganese line and between two iron lines. The presence of manganese should be ascertained from Mn-4823 before attempting to determine rhodium. The appearance of a strong line between Fe-5586 and Fe-5615, if properly spaced, will indicate rhodium. Equally strong and more persistent lines are to be found in the ultraviolet region.

Rhodium and palladium, adjacent members in the atomic series, were both discovered by Wollaston in 1804. Both are members of the platinum group and have about the same melting point as platinum. Rhodium, like iridium, cannot be dissolved in aqua regia. It is harder than platinum and is used for pen points.

Rhodium, placed in a graphite column three feet square, was one of the metals used as detectors to determine neutron penetration in the construction of the atomic bomb. The intensity of its activation indicated the number of hits in the bombardment. It is also used in thermocouples. It occurs in Mexico and in the placer sands of Russia.

The price of rhodium is $50 per ounce. Another firm lists it at $8.50 per gram.

RUBIDIUM RB (AT. WT. 85.48, AT. NO. 37)

PR(9000)	Rb 7947.6		(600)	Rb 5724.4
PR(9000)	Rb 7800.2			5700.2 Cu (350)
(1000)	Rb 7757.6			
				4217.5 Fe (200)
	6303.7 Ti	(200)		4216.1 Fe (200)
(300)	Rb 6299.2		PR(1000)	Rb 4215.5 Sr (300)
(1000)	Rb 6298.3			4213.6 Fe (100)
	6220.4 Ti	(100)		4210.3 Fe (300)
	6215.2 Ti	(100)		4203.9 Fe (200)
(800)	Rb 6206.3			4202.0 Fe (400)
			PR(2000)	Rb 4201.8
(600)	Rb 6070.7			4199.0 Fe (300)
	6063.1 Ba	(200)H		4198.3 Fe (250)

Persistent rubidium lines of great intensity occur in the deep red region at the very limit of visibility. Other very intense lines, better placed for observation, are the red lines Rb-6298 and Rb-6206 and the violet lines Rb-4215 and Rb-4201. The red lines are much more nearly free from interference than the violet ones and so constitute better identification lines.

Rubidium is one of the alkali metals, and like the other members of its group, has spectral lines more brilliant than those of any other group. All of the alkali metals have comparatively few lines, but what they lack in number they make up in brilliance. The following table shows the comparative brilliance of their strongest lines.

TABLE 17
The Alkali Metals

Element	Atomic Number	Wave-length Å	Intensity
Lithium	3	6707	(3000)
Sodium	11	5889	(9000)
Potassium	19	7664	(9000)
Rubidium	37	7800	(9000)
Cesium	55	8521	(5000)
Virginium	87	(unknown)	

Rubidium and cesium were both discovered by Bunsen and Kirchhoff in 1860. They were the first to apply the spectroscope to the detection of the elements and so became the pioneers in spectroscopic analysis. They gave the elements Latin names indicating the red and blue colors, respectively, of their most brilliant lines.

Rubidium is found in lepidolite, accompanied by lithium, potassium and sometimes cesium. Lepidolite is a universal ore for the alkali metals in much the same way that cerite is for the rare earths.

The price of rubidium iodide is $26.00 per ounce. The price of the metal is $6.00 per gram.

Probably the most easily spotted ruthenium line is Ru-6923, lying about halfway between the lines of the red potassium doublet. The stronger line Ru-4554 is also dependable in the absence of barium. Weaker, but more persistent, lines are found in the ultraviolet region.

RUTHENIUM RU (AT. WT. 101.7, AT. NO. 44)

	6938.9 K	(500)		4556.1 Fe	(150)
(300) Ru	6923.2		R(1000) Ru	4554.5 Ba	(1000)
(100) Ru	6911.4 K	(300)		4552.4 Ti	(150)

	6707.8 Li	(3000)	PR(200) Ru	3498.9
	6693.8 Ba	(600)	PR(300) Ru	3436.7
(300) Ru	6690.0			

Ruthenium was discovered by Claus in 1845, nearly half a century later than the other metals of its group. (See palladium.) It is one of the platinum metals, with the same characteristics of difficult fusibility and insolubility as the others.

The price of ruthenium chloride is $3.50 per gram; the metal $2.75 per gram.

SAMARIUM SM (AT. WT. 150.43, AT. NO. 62)

(500) Sm 6861.0		P(200) Sm 4434.3 Ca (150)
(800) Sm 6860.9		Ti (100)
		(200) Sm 4433.8 Fe (150)
(400) Sm 6734.8		4430.6 Fe (200)
(400) Sm 6734.0		4427.3 Fe (500)
(500) Sm 6731.8		4425.4 Ca (100)
	6707.8 Li (3000)R	P(300) Sm 4424.3
		4422.5 Fe (300)
(400) Sm 6589.7		(150) Sm 4421.1
(200) Sm 6570.6		(200) Sm 4420.5 Os (400)
(500) Sm 6569.3		

The most persistent samarium lines, Sm-4424-4434, are in a blue group containing samarium, iron and calcium lines; this makes quick identification difficult. There are three red groups of more intense and more readily recognized lines, as indicated above. Samarium is surpassed by few elements in its multitude of spectral lines. Probably over 4000 have been charted. It is one of the four rare earths discovered by Boisbaudran from

1875 to 1907, and is, indeed, rare; it occurs in minute proportions in cerite and has been spectroscopically observed in samarskite.

The price of samarium oxide is $22.60 per gram, but it may sometimes be purchased for less.

SCANDIUM Sc (AT. WT. 45.1, AT. NO. 21)

	5709.3 Fe (100)	P(100) Sc	4023.6
R(400) Sc	5700.2 Cu (350)		4022.6 Cu (400)
	5688.2 Na (300)		4021.8 Fe (200)
(200) Sc	5685.8	P(50) Sc	4020.3 Co (500)
	5675.7 Na (150)H	P(150) Sc	3911.8 Pr (20)
H(300) Sc	5671.8	P(125) Sc	3907.4 Gd (100)
	5670.1 Na (100)H		

In the absence of copper, scandium is probably most easily determined by its strongest line Sc-5700. The doublet Sc-5671-5686, within a sodium triplet, is easily located. The most persistent violet lines are in a crowded field. Scandium's 500 lines are fewer than those of most of the rare earths. Besides the lines listed above, certain of its compounds give bands with heads at 6072 and 6079 angstroms.

Scandium, discovered spectroscopically by Nilson in 1879, is the lightest metal having the spectrum characteristics of the heavier ones, i.e., many bright but not too bright lines in all colors. Its multiplets are numerous and its series inconspicuous. It is found in euxenite and is perhaps even scarcer than samarium.

SELENIUM SE (AT. WT. 78.96, AT. NO. 34)

P(500) Se 4742.2 Discharge Tube
P(800) Se 4739.0 Discharge Tube
P(1000) Se 4730.7 Discharge Tube
P(1000) Se 2039.8 Discharge Tube

Selenium may be excited either in a tube or by a spark; it has no arc spectrum. The element tints the arc flame blue, as may be seen by projecting the image of the arc upon a white card by means of a lens.

Selenium was discovered by Berzelius in 1817. Both it and sulfur have low melting points; both burn with bluish flames and disagreeable odors. They are found together in nature, and resemble each other chemically. Selenium cells, which are photoelectrically activated, have a variety of uses.

The price of selenium is about $0.40 per ounce.

SILICON Si (At. Wt. 28.06, At. No. 14)

	5953.1	Ti	(150)	P(400)	Si 2528.5
(50)	Si 5948.5			(400)	Si 2524.1
	5941.7	Ti	(100)	(300)	Si 2519.2
				P(500)	Si 2516.1
(500)	Si 2881.5			P(300)	Si 2506.8

Silicon may be instantaneously recognized by the ultraviolet quintuplet, Si-2506-2516-2519-2524-2528, containing two persistent lines. The visible lines are all weak, but Si-5948, between two titanium lines, is easily spotted in rocks having a fairly high silicon content, and most rocks do. In testing powdered ore for silicon one will have to use a substitute for the quartz fragment upon which the powder is burned, otherwise the silicon of the quartz will contaminate the spectrum. A large pile of powder which does not melt down to the fragment will sometimes suffice; or a table of carbon or alumina may be used. There are carbon bands, in the region of Si-5948, which wax and wane alternately with the silicon line. Since, however, the silicon line dominates the carbon band one soon learns to distinguish between them.

The ancients were familiar with the glassy mass produced by melting rocks, and certain characteristics of silicon were observed in the Middle Ages, but the discovery of the free element is attributable to Berzelius in 1823. This renowned Swedish chemist discovered four or five other elements and invented the system for chemical symbols. Davy, also, made contributions to the understanding of silicon.

Silicon makes up over one-fourth of the crust of the earth, taking second place only to oxygen. It is a non-metallic element

Characteristic Lines of the Elements

which combines with oxygen to form quartz, one of the commonest minerals. There are hundreds of silicates—more than of any other mineral group. The feldspars, pyroxenes, amphiboles, nephelites, garnets, zircons and micas are examples. Glass, cement, ceramics and many other commercial products contain silica as an essential ingredient.

A good grade of silica brings about \$30 per ton, chemically pure silicon \$0.15 per pound.

Silver Ag (At. Wt. 107.88, At. No. 47)

H(500)	Ag 5471.5		5210.3 Ti	(200)
R(1000)	Ag 5465.4	P(1500)	Ag 5209.0	
	5463.2 Fe (100)		5208.4 Cr	(500)
	5455.6 Fe (300)		5206.0 Cr	(500)
	5446.9 Fe (300)		5204.5 Cr	(400)
	4057.8 Pb (2000)R	R(1000)	Ag 3382.8	
R(800)	Ag 4055.8	PR(2000)	Ag 3280.6	

The most useful visible silver line, Ag-5465, forms an evenly spaced triplet with the two strong lines Fe-5446-5455. When the percentage of silver in an ore is small, the silver lines will often be represented only by bright flashes. The more intense line, Ag-5209, is persistent, but is so close to iron, titanium and chromium lines that more care must be exercised in using it. The most intense and persistent line, Ag-3280, is in the ultraviolet region. Ag-4055 and Pb-4057 are violet neighbors and Cu-4062 is close to them. Either lead or copper, added to the sample in the arc, will spot the silver line.

Silver has been used for ornaments and as a medium of exchange since ancient times; it was also valued by the alchemists. It was sometimes found native and often in easily manipulated ores. Silver is soft, malleable and ductile; it is the best conductor of electricity, and, next to gold, the best conductor of heat. It is used for tableware, mirrors, coinage and indelible ink and is indispensable in photography.

The silver mines of Mexico are the richest in the world, although those of Nevada surpassed them for a time in production. The chief ores are argentite and proustite.

In 1943 the price of silver was $0.71 per ounce.

SODIUM NA (AT. WT. 22.997, AT. NO. 11)

(500) Na 6160.7	P(300) Na 5688.2
(500) Na 6154.2	P(80) Na 5682.6
PR(5000) Na 5895.9	P(300) Na 3302.9
PR(9000) Na 5889.9	P(600) Na 3302.3

There is never any doubt about the identification of sodium. The doublet Na-5889-5895 is present in almost every spectrum, for carbons and other electrodes are almost sure to carry some traces of it. So omnipresent are the lines that it is customary to set the spectroscope scale by them. The sodium spectrum consists almost completely of doublets whose components draw closer and closer together as their wave-lengths decrease. Over sixty spectral lines of other elements have been observed in the six-angstrom space which separates the lines of the principal sodium doublet.

Sodium was first isolated by Davy in 1807 by electrolysis.

Sodium nitrate occurs as a natural deposit in Chile, Peru and Bolivia. The oligoclase feldspars and albite also contain sodium, as do the sea and sea plants. Sodium carbonate has wide commercial application in the manufacture of glass and paper, caustic soda, soap, cleaning compounds and chemicals. The bicarbonate is used in baking powders, the thiasulfate in photography, the silicate in glass and water-glass, and the vapor for brilliant incandescent lights on bridges.

Lithium, sodium, potassium and rubidium are all alkali metals having a single electron in their outer shells and having very similar spectra. Sodium, in its metallic state, quickly oxidizes in the air, and reacts violently with water.

The price of sodium metal is $0.85 per pound.

Characteristic Lines of the Elements

Strontium Sr (At. Wt. 87.63, At. No. 38)

4611.2 Fe (200)	(1000) Sr 7070.1
4607.6 Fe (50)	7059.9 Ba (2000)
PR(1000) Sr 4607.3	
4605.3 Mn (150)	P(400) Sr 4077.7
4602.9 Fe (300)	P(200) Sr 4832.0
4602.8 Li (800)	P(200) Sr 3464.5

Traces of strontium can be detected by the very sensitive and persistent line Sr-4607, so commonly seen in the spectrum of ordinary rocks. Neighboring iron, manganese, and lithium lines serve as spotters. If considerable strontium is present the doublet Sr-6878-6892, about ten angstroms from the doublet Ba-6865-6867 will serve as a check. The lines at 7070, 4077, 4832 and 3464 angstroms may all be useful on occasion.

Strontium, barium and calcium belong to the same chemical group; all have comparatively few lines and strontium has only 200, about half as many as the others. In the 110-volt arc the spectrum is very brilliant and ranges through all the colors.

Strontium was discovered by Davy by the electrolytic method in 1808. It is one of the elements originally termed *alkaline earths* because all the forms known then were earthy as compared with the metals. Davy showed that all were really metallic, chemically very active, and electropositive. Strontium is used in large quantities in sugar refining and in fireworks.

Strontium is a yellow metal with a low melting point. Its compounds impart a brilliant red color to flames and are used in Roman candles and other pyrotechnics. Celestite is the most common ore.

The price of strontium nitrate is $0.43 per ounce.

Sulfur S (At. Wt. 32.06, At. No. 16)

P(2000) S 9323.4 Discharge Tube
(700) S 5606.1 Discharge Tube
P(500) S 4694.1 Discharge Tube

Sulfur gives no arc lines but has strong visible and infrared Geissler-tube spectra. In the arc, unmistakable, acrid sulfur dioxide fumes are given off by many ores. Moreover, the chemical determination of sulfur is very simple. A mixture of sodium bicarbonate (baking soda) and the pulverized mineral or sample is fused in the arc; then the bead is placed upon a clean silver coin and moistened with a drop of water. A black or yellow stain on the coin instantly betrays the presence of even a trace of sulfur.

Sulfur was known to the ancients as brimstone, and on account of its inflammable nature was regarded by the alchemists as proof of the transmutation of the metals. It is often found near volcanoes, both in solid form and as fumes issuing from vents.

Sulfur has more allotropic forms than any other element, since it may be rhombically or monoclinically crystalline, soft and soluble, liquid, plastic, amorphous, or colloidal, with more or less peculiar characteristics for each form. It combines with the other elements to form a multitude of compounds and ores. It is used in insecticides, matches, bleaching agents, gunpowder, and in the manufacture of sulfuric acid. Sulfur is taken from deep wells in the southern states.

A recent price was $18.00 per ton.

Tantalum Ta (At. Wt. 180.88, At. No. 73)

(150)	Ta 6928.5		4674.7 Cu (200)
(150)	Ta 6927.3	(300)	Ta 4669.1
			4668.5 Na (200)
	6494.9 Fe (400)		4668.1 Fe (125)
	6469.9 Ba (800)		4667.4 Fe (150)
(500)	Ta 6485.3	(300)	Ta 4661.1
	6482.9 Ba (100)		
			4554.0 Ba (1000)
(100)	Ta 4832.1	(200)	Ta 4553.6
(150)	Ta 4825.4		4552.4 Ti (150)
(100)	Ta 4819.5	(400)	Ta 4551.9
(150)	Ta 4812.7	P(125)	Ta 3318.8
	4810.5 Zn (400)	HP(300)	Ta 3311.1

The intense doublet, Ta-4661-4669 may be spotted near Cu-4674. A less intense but more striking quadruplet Ta-4812-4819-

4825-4832 lies only two angstroms from the intense blue line Zn-4810. The persistent lines are in the ultraviolet region.

Tantalum was discovered by Ekeberg in 1802, but it had very little importance until the day of electronics, synthetic rubber, and plastic surgery more than a century later. During World War II it had top priority rating. Ore deposits occur in the pegmatites of the Black Hills and in the interior of Brazil. Three thousand tons of rock are sometimes hand picked to obtain a single ton of ore. During the war, Brazil made its entire tantalite output available to the United States and established an identification laboratory in Paraiba, the center of the mining district.

Tantalum metal may be drawn into fine wire, highly resistant to heat, for use as electric-bulb filaments and in radio-tube grids. It resists corrosion and attack by the strongest acids—even aqua regia—and does not oxidize. Tantalum is used in surgery since it does not irritate tissue, cause adhesions, nor accumulations of serum. It is used in the form of screws, nails, and metal plates for mending fractures, and as foil for wrapping nerves and tendons to keep scar tissue from adhering to them. It may be used to replace bone loss in the skull and for artificial eyeballs. In the form of wire, it is used in suturing nerves and tendons, but is somewhat brittle. More pliable alloys are being sought.

The price of tantalum powder is $0.50 per gram.

Tellurium Te (At. Wt. 127.61, At. No. 52)

(250) Te 5708.0 Discharge Tube   P(600) Te 2385.7 Arc
(250) Te 5649.3 Discharge Tube   P(500) Te 2383.2 Arc

In the arc, tellurium yields no visible lines, but fairly intense ultraviolet ones. The yellow lines Te-5649-5708 may be seen in either the spark or the Geissler Tube.

Tellurium was discovered in gold ore by Reichenstein in 1782. It is a silvery, brittle element resembling sulfur and selenium in chemical properties, and like sulfur, very undesirable in metallic alloys. It is one of the few elements naturally uniting with gold. Tellurides of both gold and lead are found in Australia and

California. It also occurs native in Colorado. Its hydrogen compounds have a very vile odor and if taken into the human system taint the breath like garlic. It is very toxic.

The price of Tellurium is $2.50 per ounce.

TERBIUM TB (AT. WT. 159.2, AT. NO. 65)

	4282.4 Fe (600)		(125) Tb	4033.0 Ga	(1000)
	4280.5 Gd (200)			Mn	(400)
(200) Tb	4278.5 Ti (50)		(50) Tb	4031.6 La	(400)
H(50) Tb	4276.7 Ti (50)			4030.7 Mn	(500)
	4274.8 Cr (4000)				
	4271.7 Fe (1000)		P(200) Tb	3561.7	
			P(200) Tb	3509.1	

Terbium may be recognized by its most intense visible doublet, Tb-4276-4278, or by the Tb-4031-4033 doublet. There is also a doublet in the ultraviolet region as indicated. It has 2600 other weak spectral lines.

Terbium was discovered in 1843 by Mosander, the Swedish chemist who, four years earlier, had discovered cerium and lanthanum. In 1868, L. Smith believed he had discovered a new earth in an ore from Carolina and named it mosandrium in honor of the discoverer of the first rare earths. It was later found to be an intimate mixture of terbium with half a dozen other elements rather than a new one.

Terbium has been separated from the other rare earths by the fractional crystallization of the bromates. The oxide is black, the salts blue; the metal has never been isolated.

The price of the oxide is $1.20 per milligram.

THALLIUM TL (AT. WT. 204.39, AT. NO. 81)

	6556.0 Ti (150)		5351.0 Ti (50)
	6554.2 Ti (125)	PR(5000) Tl	5350.4
(200) Tl	6552.6		5349.4 Ca (12)
(300) Tl	6549.7		
	6546.2 Fe (150)	PR(3000) Tl	3775.7
		P(2000) Tl	3519.2

A few of thallium's 300 lines are intense—especially the bright green one Tl-5350. Unfortunately, there is a calcium line only an angstrom away, which, in the 110-volt a-c arc is more intense than indicated in the literature and which must be watched. The doublet Tl-6549-6552 will show up whenever the element is present in quantity. There are also strong and persistent lines in the ultraviolet region.

Thallium was discovered spectroscopically in 1861 by Crookes. who, the following year, succeeded in isolating it completely. In 1863 he was elected to the Royal Society, in 1897 knighted, and in 1907 awarded the Nobel prize in chemistry. He may be credited with the discovery of electrons, since he detected them streaming from the cathodes of his highly evacuated tubes. His experiments paved the way for X-rays and a new conception of the structure of the atom.

Thallium is one of the by-products in the manufacture of sulfuric acid. Like selenium, it collects in the flues during the reduction of iron pyrites and is later recovered. It is a little softer than lead and similar to it in malleability and low melting point. It forms two series of compounds, in one of which it is univalent and the other trivalent. The compounds are extremely toxic, causing symptoms like those of lead poisoning. It is used in optical glass and in rodent poisons.

The price is $0.35 per gram for the pure metal.

THORIUM TH (AT. WT. 232.12, AT. NO. 90)

	5018.4 Fe			4868.2 Ti (100)	
(50)	Th 5017.2		(20)	Th 4863.1	
	5016.1 Ti (100)			4859.7 Fe (150)	
(10)	Th 5015.8 Fe (500)		(10)	Th 4858.3	
	5014.2 Ti (100)		(12)	Th 4850.4	
(25)	Th 4282.0		P(8)	Th 4019.1	
(20)	Th 4281.0		P(8)	Th 3601.0	
(20)	Th 4277.3				

Thorium has over 2500 spectral lines, but not a single one of even medium intensity. Moreover, the most intense lines occur in masses of iron and titanium lines. Probably the best chance of identifying the element is by using the triplet Th-4277-4281-4282, or the group Th-4850-4863. Th-4019 is a very weak but persistent line. The ultraviolet lines are equally weak.

Thorium was discovered by Berzelius in 1828 and named after the Scandinavian god of thunder. The name was more appropriate than the Swedish chemist probably realized, for the element is capable of fission and is more plentiful than uranium. The only highly radioactive natural elements are uranium, thorium, radium and actinium. All of them spontaneously emit alpha particles, and so lower their atomic numbers and change to other elements. The new forms are unstable, some have very short lives, some very long. All finally become stable in lead, which is the final product. Thorium fission is produced only by bombardment with high-speed neutrons; it then splits into nearly equal parts and produces two of the elements near the middle of the atomic series.

Thorium is present in the earth to the extent of about 12 parts per million and is widely distributed in monazite sands, especially in Brazil and India. It is also found in the black sands of California, Idaho, Florida and Carolina. It will burn in air; thoria, the oxide, is used in the radiant elements of gas mantles. Thorium is less radioactive than radium.

A recent price of thorium is $5.75 per pound.

Thulium Tm (At. Wt. 169.4, At. No. 69)

(300) Tm 6604.9 Sm (200)   (300) Tm 4359.9 Pr (170)
                                       4358.3 Hg (3000)
       6462.5 Ca (125)
(400) Tm 6460.2   4247.4 Fe (200)
       6457.9 Eu (500)   (500) Tm 4242.1
       6450.8 Ba (100)   4238.8 Fe (200)

P(250) Tm 3761.9

Thulium may be recognized by its strongest violet line, Tm-4242, near the middle of an iron doublet. In the red-orange field Tm-6460, near Ca-6462, is also placed well for observation.

Characteristic Lines of the Elements

Thulium and scandium were both discovered by Cleve in 1879. James separated the sesquioxide in 1911. Its different salts are of several colors, green, white and blue. Its solutions give both bright line and absorption spectra. Thulium is among the rarest of the rare earths, but occurs along with the others in such minerals as gadolinite, fergusonite, polycrase and euxenite. Thulium belongs to the erbium family.

TIN SN (AT. WT. 118.7, AT. NO. 50)

	5635.5 C	Band	4528.6	Fe	(600)
(50) Sn	5631.6		4527.3	Ti	(100)
	5629.3 Ti	(30)	4526.9	Ca	(100)
	5624.5 Fe	(150)	4525.1	Fe	(100)
	5615.6 Fe	(400)	HP(500) Sn 4524.7		
			4523.3	Mn	(50)
PR(300) Sn	2839.9		4523.2	Ba	(60)
HP(500) Sn	3175.0		4522.8	Ti	(100)

Tin, in any considerable amount in an ore, is frequently more easily recognized by Sn-5631 than by its more intense lines because of its freedom from ordinary interference. It also falls within a carbon band, only four angstroms from its head, and thus can be accurately placed. The strongest visible line Sn-4524 is persistent and wide. It is characterized by a fuzziness which also distinguishes it from its iron and barium neighbors. Mere traces of tin may be obscured by Fe-4525, but if, by comparison, line Sn-4524 should equal or surpass Fe-4528 in brilliance, the presence of tin would be indicated. There are ultraviolet lines of equal intensity and persistence as shown.

Tin was known to the ancient Hebrews; implements of its alloys have been found by archaeologists. It was one of the articles of trade of the seafaring Phoenicians, and the Romans secured tin from the mines of Cornwall.

Tin ore may be smelted by mixing it with one-sixth of its weight of powdered charcoal and a small quantity of fluorspar as a flux for the silica usually present. Roasting continues for six hours; the slag is then raked off and the tin found at the bottom of the receptacle. Cassiterite, the chief ore, is mined in the East Indies, and also in Bolivia. Tin does not oxidize in air and its chief use is as a coating for iron and copper. It is used in solders.

The 1944 price of tin was $0.52 per pound.

TITANIUM TI (AT. WT. 47.9, AT. NO. 22)

(400)	Ti 6146.2		5014.9	Fe (500)
	6141.7 Ba (2000)	(140)	Ti 5014.2	
			5012.0	Fe (300)
(100)	Ti 4759.2	(200)	Ti 5007.2	
(125)	Ti 4758.1		5006.1	Fe (500)
	4754.0 Mn (400)	P(200)	Ti 4999.5	Fe (300)
			4994.1	Fe (200)
P(500)	Ti 3653.4	P(200)	Ti 4991.0	Fe (80)
P(300)	Ti 3642.6		4983.8	Fe (200)
		P(300)	Ti 4981.7	Fe (200)

Titanium is quickly recognized by an outstanding quintuplet with its mid-line a fraction of an angstrom from 5000 Å. The lines of the group, about equally spaced and of similar brilliance, are recognizable even in low-dispersion instruments. Intense iron lines sometimes crowd into the group and disorganize the picture when only a trace of titanium is present. The most intense visible line is Ti-6146; the most intense and persistent ultraviolet line is Ti-3653. Titanium doublets abound and frequently assist in placing the lines of other elements, e.g., Ti-4758-4759 in the manganese group. Only iron and calcium have more numerous lines in the average common rock than titanium. It is always present in clays and volcanic rocks, and even though the quantity is small, the lines may be surprisingly clear. Once considered a rare element, it is now recognized as being one of the most widely distributed.

Titanium was discovered by Gregor in 1789 and the pure metal was first prepared by Hunter in 1910. It is a silvery white metal, burns in air and quite unlike other metals, also in nitrogen. The metal gives tensile strength and hardness to steel. The oxide makes a white pigment. The chief ores are the rutile of

Virginia and the ilmenite of California. Titanite is found in Europe.

The price of the refined metal is $5.00 per pound.

TUNGSTEN W (AT. WT. 183.92, AT. NO. 74)

(20)	W 5055.5		(150)	W 4680.5 Zn	(300)
(25)	W 5054.6 Ba	(12)	(200)	W 4659.8	
(60)	W 5053.2			4656.4 Ti	(150)
	5052.8 Ti	(50)		4651.1 Cu	(250)
	5051.6 Fe	(200)			
	5049.8 Fe	(400)	P(60)	W 4302.1 Ca	(50)
			P(70)	W 4294.0 Fe	(700)
	4844.3 Mn	(80)	P(45)	W 4008.7 Ti	(130)
(50)	W 4843.8				
	4840.8 Ti	(125)			

Although tungsten has more spectral lines than any other of the elements except cerium, iron and uranium, and although its spectrum shows distinctive groups, it is still not a very easy element to identify because the lines are weak. Perhaps W-4659 and W-4680 are the most satisfactory lines for its determination, although W-4843 has a convenient titanium spotter. The close triplet W-5053-5054-5055 is weak but determinative. There are no strong ultraviolet lines.

Tungsten was discovered by the d'Elhujar brothers in 1783. It has the highest melting point of all the metals and so makes long-lasting and brilliant filaments for electric lamps. Although it is not ductile, it can be rolled and pressed into wire. It is self-hardening and crystalline. It may be dissolved only in a combination of nitric and hydrofluoric acid. Its compounds are used as mordants and for fireproofing.

An alloy of 8 to 20 per cent of tungsten and 3 to 5 per cent of chromium in steel is used for tools for high-speed turning; these tools may become red hot without losing their temper. Different metals impart different qualities to steel: manganese, hardness and ability to hold temper; chromium, tensile strength and elasticity; nickel, elasticity and resistance to corrosion; tungsten, durability for high-speed tools. The chief ores are scheelite and wolframite.

The wartime price of tungsten was $3 per pound.

URANIUM U (AT. WT. 238.07 AT. No. 92)

(100) U 6449.1
        6400.0 Fe (200)
(100) U 6395.4
        6393.6 Fe (100)

        5918.5 Ti (80)
(125) U 5915.3 Co (200)
        5899.3 Ti (150)

(60) U 5492.9 W (50)
(12) U 5482.5
(30) U 5481.2 Mn (50)
(15) U 5480.2 Sr (100)
      5474.9 Fe (100)
      5473.9 Fe (100)

Uranium may usually be identified by its most intense line, U-5915, unless the sodium doublet is too intense. The orange line, Ti-5918, must not, however, be taken for uranium. U-5492 is also of value and is usually clear of interference. The ultraviolet spectrum is weak. Uranium has over 5000 lines of intensity greater than (1) and there is not a brilliant one among them all. Only cerium has more numerous spectral lines.

Uranium was discovered in 1789 by Klaproth, a Berlin chemist, who recognized the importance of Lavoisier's work and imitated his thoroughness. He named the new element after the planet discovered by Herschel eight years earlier.

Uranium is the radioactive source element used in the atomic bomb. Of its several isotopes U-235 forms about 1 part in 140, and is capable of setting up an explosive chain reaction. Isotope U-238 absorbs slow-speed neutrons, changes in succession to neptunium and plutonium and is then also ready for the chain reaction. Thus uranium, heretofore used chiefly for coloring ceramics, has become the greatest potential source of power ever placed in the hands of man.

Uranium occurs chiefly in pitchblende and in carnotite. The largest deposits are in Canada, the Belgian Congo, and in Colorado. The most sensitive means of detection is the Geiger-Müller counter, which during the war were produced in numbers and used at Great Bear Lake in the location of ore bodies. The amount of uranium present in a mineral in comparison with the quantity of lead, affords a means for estimating the age of the rock.

The prewar price of uranium nitrate was $5.00 per pound; it is now $6.23 per pound.

VANADIUM V (AT. WT. 50.95, AT. No. 23)

	5709.3	Fe (100)	PR(80)	V 4389.9	
(200)	V 5706.9		PR(125)	V 4384.7	
(200)	V 5703.5			4383.5 Fe	(1000)
	5700.2	Cu (350)		4381.7 Mo	(150)
(300)	V 5698.5		PR(200)	V 4379.2	
	5688.2	Na (300)		4378.2 Cu	(200)H
	5675.7	Na (150)		4375.9 Fe	(500)
(200)	V 5627.6		(300)	V 6077.3	
(150)	V 5626.0				
(170)	V 5624.5	Fe (150)	PR(500)	V 3185.3	
	5615.6	Fe (400)	PR(500)	V 3183.9	

Vanadium, even when only in traces, may be quickly identified by V-4379, its most persistent visible line. If copper or molybdenum is present in the sample, however, the spacing from Fe-4383 and Fe-4375 must be carefully judged. When there is more vanadium than iron in the rock, V-4379 will be seen as the end line of a beautiful blue vanadium quintuplet with regular spacing. Note the triplets V-5624-5626-5627 and 5698-5703-5706, and the persistent ultraviolet doublet V-3183-3185. Over 3000 vanadium lines have been charted.

The persistent vanadium lines are very common in ordinary rocks and minerals, but the element, although so widespread, is usually present only in traces. Vanadium lines are so sensitive that they may be detected when the metal is present in concentrations of only 0.0001 per cent, and so even rather strong lines must not be interpreted as indicating a high concentration.

The discovery of vanadium is attributed to the Swedish scientist Sefström in 1830, and the first separation of the metal to

Roscoe in 1869. Vanadinite was first found in Mexico, but Peru is the chief source of vanadium ore. The carnotite of Colorado and Idaho contains vanadium, radium and uranium. Vanadium is a steel-hardener and the chloride a mordant.

The ore is valued according to its content of vanadium oxide, the price being $0.27 for each pound of oxide present. The metal is listed as $2.25 for 100 grams.

XENON XE (AT. WT. 131.3, AT. NO. 54)

P(2000) Xe 4671.2 Discharge Tube
P(1000) Xe 4624.2 Discharge Tube
(500) Xe 4500.9 Discharge Tube

Xenon is a heavy gas constituting only one part in millions of the atmosphere; it is obtained by first liquefying air under intense pressure and low temperature, and then evaporating off the other ingredients until only the xenon remains. Although it is an inert gas chemically, it yields a brilliant spectrum in the Geissler tube.

The spectrum of xenon resembles that of argon and those of the other inert gases. It has over 1000 lines, many of them brilliant. It was discovered by Ramsay and Travers in 1898.

The price of xenon is $600.00 per liter.

YTTERBIUM YB (AT. WT. 173.04, AT. NO. 70)

(1000) Yb 6799.6		5572.8 Fe (300)
6791.0 Sr (200)		5569.6 Fe (300)
	(1500) Yb 5556.4	
6675.2 Ba (500)		
(1000) Yb 6667.8	PR(1000) Yb 3887.9	
	PR(500) Yb 3694.2	

Ytterbium may be detected by any one of the intense lines above if it can be found in sufficient concentration. The most intense line, Yb-5556, is visible; the most persistent, Yb-3887, is in the ultraviolet region. Ytterbium was discovered by Marignac in 1878. It is another of the extremely rare earths found associated

Characteristic Lines of the Elements

Yttrium Y or Yt (At. Wt. 88.92, At. No. 39)

(200) Yt 6132.1 Band		4678.1 Cd (200)H
(200) Yt 6003.6 Band	P(80) Yt 4674.8 Er (50)	
(300) Yt 5987.6 Band		4672.0 Pr (100)
(600) Yt 5973.0 Band	P(50) Yt 4643.6 Pr (60)	
		4639.3 Ti (140)
6438.4 Cd (2000)		
(150) Yt 6435.0 Sm (25)		4375.9 Fe (500)
P(100) Yt 3600.7	(150) Yt 4374.9 Er (40)	
		4373.8 Gd (200)

The determination of yttrium is not too easy since the persistent and intense visible lines are in a field crowded with the lines of the other rare earths. More reliance may often be placed upon Y-6435 and upon the strong bands at the red end of the spectrum, which show up well in yttrium compounds.

Yttrium, discovered by Gadolin in 1794, is a rare element resembling the rare earths and accompanying them in such minerals as gadolinite, euxenite and polycrase. It has been isolated by electrolytic separation, yielding a gray lustrous metal.

Yttrium and scandium have strikingly similar spectra. Yttrium, when ionized by the loss of electrons in the outer shell has a simpler spectrum than the neutral element. Doubly ionized, 2 electrons driven off, it shows still fewer and stronger lines like an alkali metal. Yttrium oxide is listed at $3.25 for 25 grams.

Zinc Zn (At. Wt. 65.38, At. No. 30)

4821.0 Fe (200)	
P(400) Zn 4810.5	PH(1000) Zn 6362.3
4722.6 Bi (1000)	
P(400) Zn 4722.1	P(800) Zn 3345.0
4684.8 C Band	P(800) Zn 3302.5
4681.9 C Band	
P(300) Zn 4680.1	PR(800) Zn 2138.5
4674.7 Cu (200)	

Although the zinc spectrum has fewer lines than those of most of the metals, it is among the most easily recognized. Its brightest line, Zn-6362, stands out clearly in the red, and its blue triplet Zn-4680-4722-4810 shows up clearly even in low-dispersion instruments. The less brilliant, but more persistent, Zn-2138, lies in the ultraviolet region.

Brass, the alloy of zinc and copper, is often mentioned in the Old Testament along with silver and gold. There were offerings of brass, rings of brass, staves, altars, serpents, plates, pots, cymbals, shields, gates and images of brass. Samson's fetters and Goliath's helmet were of brass. Coins of brass containing zinc, dating back nearly to the first century A.D., have been found but probably isolated zinc was not known to the ancients. It was described by Paracelsus, the Swiss alchemist in 1520, but as late as 1675 its characteristics were still uncertain as it was even then confused with bismuth.

Zinc, although not too widely distributed, is a common metal with numerous uses. It is used in galvanizing iron, as sheet metal, and for photo-engraving. Zinc oxide is probably the best of all paint hardeners. Sphalerite and smithsonite are among its important ores.

The price of zinc is sometimes as low as $0.05 per pound, but the war-time price was $0.08 or more.

ZIRCONIUM ZR (AT. WT. 91.22, AT. NO. 40)

```
            6146.2 Ti  (400)       P(100)  Zr 4772.3
  (300) Zr  6143.2                 P(100)  Zr 4739.4 Mn (150)
            6141.7 Ba (2000)               4722.1 Zn (400)
            6137.6 Fe  (100)       P(60)   Zr 4710.0 Ti (100)
            6136.6 Fe  (100)         (50)  Zr 4688.4
  (300) Zr  6134.5                 P(125)  Zr 4687.8
  (500) Zr  6127.4 Cu  (80)                4681.9 Ti (200)
            6126.2 Ti  (150)               4680.1 Zn (300)H
            6122.2 Ca  (100)
                                   P(400)  Zr 3601.1
```

Zirconium may be easily detected by the red triplet Zr-6127-6134-6143, the blue triplet Zr-4687-4688-4710, or by its ultraviolet lines. Zirconium lines are seldom seen in ordinary rocks, since none of them are brilliant, and the element is not too abundant.

Pure zirconium has been produced, but is uncommon. It was first found in 1789 by Klaproth, the discoverer of uranium. Berzelius isolated it in 1824.

Many crystals and minerals when viewed under violet or ultraviolet light glow brilliantly in unexpected colors. Scheelite, willemite, and fluorite are among the best examples of such excitation. Many zircons (zirconium silicate crystals) show an orange fluorescence. The activity may be due to the presence of the hafnium, always found with zirconium, or perhaps to the zirconium itself. Zircon crystals, when of good quality and size, have gem value. Zircon sands are frequent along the Pacific coast, and, being heavier than quartz sands, may be panned out like gold. Much zirconium is also imported from Brazil. It is used in enameling, in spot-welding electrodes, radio tubes, photoflash bulbs, gas mantles, paint, and as an abrasive.

The price of zirconium oxide is about $0.50 per pound.

OUTSTANDING ARC LINES WITH INTENSITIES
FOR AN ALPHABETICAL LIST OF ELEMENTS

Aluminum	8572.6 200	2369.6 40	4934.0 400
Al	7924.6 300	2349.8 250	4554.0 1000
8874.5 100	7844.4 100	2288.1 250	3501.1 1000
7836.8 15	4033.5 70	2271.3 25	
6698.7 10	3267.5 150	2165.5 50	Beryllium
6696.3 10	3232.4 150	2144.1 50	Be
3961.5 3000	3029.8 100	2134.8 18	4572.6 15
3944.0 2000	2877.9 250	2113.0 50	3321.3 1000
3092.7 1000	2851.1 50	2031.4 75	3321.0 100
3082.1 800	2769.9 100	2003.3 300	3321.0 50
3066.1 25	2718.8 50	2002.5 20	3131.0 200
3064.3 20	2682.7 50		3130.4 200
3054.6 20	2670.6 50	Barium	2898.1 15
3050.0 18	2652.6 50	Ba	2650.7 25
2660.3 150	2598.0 200	9830.3 300	2650.7 10
2652.4 150	2528.5 300	9370.0 300	2650.6 25
2575.1 200	2311.4 150	8559.9 400	2650.6 20
2567.9 200	2175.8 300	7905.7 300	2650.5 30
2373.3 200	2068.3 300	7672.0 400	2650.4 100
2373.1 100		7392.4 400	2494.7 30
2367.0 150	Arsenic	7280.2 1000	2494.5 30
2321.5 5	As	7120.3 800	2494.5 25
2269.2 15	8869.9 100	7059.9 2000	2350.8 12
2263.4 60	8821.7 150	6865.7 200	2350.6 25
2174.0 8	8564.7 100	6693.8 600	2348.6 2000
2168.8 8	8541.6 50	6675.2 500	2175.0 25
2129.4 5	8428.9 100	6595.3 1000	2084.0 2
	3119.6 100	6496.9 800	2056.5 100
Antimony	3075.3 60	6141.7 2000	2033.3 10
Sb	3032.8 125	6110.7 200	
9949.1 400	2898.7 25	6063.1 200	Bismuth
9578.6 400	2780.1 75	5997.0 150	Bi
9518.6 400	2492.9 25	5853.6 300	9657.2 2000
8711.1 50	2456.5 100	5777.6 500	9342.5 500
8682.7 100	2381.1 75	5535.5 1000	8907.9 200
8619.5 150	2370.7 50	5519.1 200	8761.5 100

Characteristic Lines of the Elements

Bismuth (cont.)					
8628.0	100	2088.9	15	8078.9	100
7838.7	400	Bands:		8015.7	200
5552.3	500	2850.6	50	7990.6	100
4722.5	1000	2675.3	60	7944.1	800
4121.5	125	2588.0	50	7609.0	500
3596.1	150	2251.4	150	7379.9	35
3510.8	200			7270.7	15
3067.7	3000	Cadmium		7229.0	35
3024.6	250	Cd		7228.5	500
2897.9	500	7385.3	800	6983.4	25
2809.6	200	7383.9	1000	6973.2	500
2780.5	200	7346.2	1000	6870.4	200
2627.9	200	6438.4	2000	6825.2	15
2400.8	200	6325.1	100	6723.2	500
2276.5	100	6111.5	100	6586.5	500
2228.2	100	6099.1	300	6212.8	100
2156.9	75	5085.8	1000	4593.1	1000
2134.3	100	4799.9	300	4555.3	2000
2133.6	100	4678.1	200	3876.3	300
2110.2	250	4075.7	500		
2061.7	300	3612.8	800	Calcium	
		3610.5	1000	Ca	
		3466.2	1000	8662.1	1000
		3403.6	800	8542.0	1000
Boron		3261.0	300	8478.0	300
B		3252.5	300	7326.1	400
3451.4	5	2980.6	1000	7148.1	500
2497.7	500	2748.5	5	6717.6	80
2496.7	300	2573.0	3	6462.5	125
2652.5	4	2312.8	1	6122.2	100
2089.5	150	2288.0	1500	6102.7	80
2088.9	100	2265.0	25	5598.4	35
Spark:		2239.8	80	5594.4	35
3451.4	100	2144.3	50	5588.7	35
2918.1	2			4585.8	125
2566.4	2	Caesium		4581.4	100
2566.2	15	Cs		4526.9	100
2557.5	6	9208.4	200	4455.8	100
2497.7	400	9172.2	1000	4454.7	200
2496.7	300	8943.5	2000	4435.6	100
2266.9	2	8761.3	500	4434.9	150
2266.3	2	8521.1	5000	4425.4	100
2089.5	20	8079.0	1000	4226.7	500

3973.7	200			
3968.4	500			
3957.0	80			
3933.6	600			
Carbon				
C				
9658.4	250			
9405.7	300			
9111.8	150			
9094.8	500			
9088.5	200			
9061.4	350			
5992.6	100			
2478.5	400			
Bands, Cn:				
5858.2	400			
5730.0	150			
5354.1	100			
5165.2	200			
3883.4	—			
3848.7	200			
2799.7	50			
2785.4	30			
2742.6	20			
2740.0	12			
2698.3	20			
2680.8	15			
2659.6	20			
2600.3	20			
2569.5	12			
2539.1	12			
Cerium				
Ce				
5696.9	40			
5669.9	50			
5522.9	100			
5512.0	50			
5409.2	50			
5353.5	50			
4528.4	30			
4527.3	50			

Cerium (cont.)

4523.0	35
4522.0	25
4486.9	40
4483.8	40
4460.2	60
4449.3	50
4429.2	35
4373.8	40
4349.7	40
4296.6	40
4263.4	40
4186.5	80
4165.6	40
4142.3	65
4042.5	50
4040.7	70
4012.3	60

Chromium Cr

5409.7	300
5345.8	300
5208.4	500
5206.0	500
5204.5	400
4936.3	200
4922.2	200
4829.3	200
4801.0	200
4789.3	300
4756.1	300
4737.3	200
4616.1	300
4580.0	300
4351.7	300
4344.5	400
4339.4	300
4337.5	500
4289.7	3000
4274.8	4000
4254.3	5000
3976.6	300
3963.6	300
3919.1	300
2780.7	600

Cobalt Co

6450.2	1000
5991.8	900
5647.2	600
5530.7	500
5353.4	500
5352.0	500
5343.3	600
5342.7	800
5341.3	300
5331.4	500
5247.9	500
4867.8	800
4813.4	1000
4663.4	700
4581.5	1000
4565.5	800
4530.9	1000
4121.3	1000
4118.7	1000
3995.3	1000
3502.2	2000
3474.0	3000
3465.8	2000
3455.2	2000
2453.5	3000

Columbium Cb

6990.3	100
6828.1	150
6677.3	200
6660.8	300
5900.6	200
5838.6	200
5671.0	200
5664.7	100
5350.7	150
5344.1	400
5320.9	500
5318.6	100
5276.1	200
5271.5	200
5219.0	100
5164.3	150
5160.3	200
5152.6	100
5124.7	200
5078.9	300
5039.0	200
4672.0	150
4100.9	300
4079.7	500
4058.9	1000

Copper Cu

8092.6	400
6920.0	100
5782.1	1000
5700.2	350
5220.0	100
5218.2	700
5153.2	600
5105.5	500
4704.5	200
4674.7	200
4651.1	250
4586.9	250
4539.6	100
4530.8	200
4509.3	150
4480.3	200
4378.2	200
4275.1	80
4075.5	40
4062.6	500
4022.6	400
3273.9	3000
3247.5	5000
2824.3	1000
2592.6	1000

Dysprosium Dy

4308.6	100
4295.0	20
4294.9	25
4256.3	25
4245.9	25
4225.1	40
4221.1	60
4218.0	50
4211.7	200
4202.2	20
4201.3	30
4198.0	20
4194.8	50
4191.6	40
4186.8	100
4167.9	50
4077.9	150
4073.1	80
4045.9	150
4000.4	400
3996.6	200
3944.6	300
3872.1	300
3757.3	200
3645.4	300

Erbium Er

5902.0	30
5456.6	30
5433.8	12
5422.8	30
5414.6	50
5395.8	50
5279.3	30
4883.6	30
4845.6	50
4527.2	50

Characteristic Lines of the Elements

Erbium (cont.)		Fluorine F Bands, CaF:				Germanium Ge	
4505.9	40	4205.0	200	4870.0	200	4685.8	20
4500.7	20	4129.7	150	4865.0	400	4260.8	20
4487.2	20			4805.8	200	4226.5	200
4481.2	20			4801.0	200	3269.4	300
4473.5	10			4781.9	200	3124.8	200
4451.5	10	6632.7	300	4743.6	300	3067.0	60
4449.7	30	6629.4	200	4732.6	300	3039.0	1000
4384.6	30	6626.1	200	4466.5	200	2754.5	30
4286.5	20	6622.9	200	4419.0	200	2740.4	10
4211.7	30	6527.6	200	4369.7	250	2709.6	30
4159.8	20	6512.0	300	4327.1	500	2691.3	25
4077.9	20	6508.7	200	4325.6	500	2664.1	4
4057.8	30	6064.4	200	4251.7	300	2651.5	30
4053.8	20	5779.4	200	3768.4	20	2651.1	40
3987.9	100	5291.0	200	3646.1	200	2592.5	20
		Tube:				2589.1	6
		6856.0	1000	Gallium Ga		2556.3	20
Europium Eu		6834.2	300			2533.2	5
		4446.5	40	4172.0	2000	2497.9	8
6864.5	1000	4299.1	150	4032.9	1000	2417.3	10
6645.1	1000	4246.1	300	2944.1	10	2379.1	3
6437.6	700	4103.5	300	2943.6	10	2338.6	2
6400.9	700	4103.0	150	2874.2	10	2327.9	3
6303.3	700	3505.7	10	2719.6	5	2314.2	3
6049.5	1000	3505.6	600	2659.8	5		
6018.1	1000	3505.5	20	2500.7	5		
5992.8	1000	3503.3	400	2418.6	3		
5967.1	2000	3501.4	200	2371.3	3		
5966.0	1000	2556.1	15	2338.5	2	Gold Au	
5831.0	2000			2338.2	3		
5818.7	1000	Gadolinium Gd		Tube:		7510.7	20
5765.2	2000			7198.7	60	6278.1	700
5645.8	1000	7168.4	500	7000.4	70	5956.9	35
5547.4	1000	6996.7	200	2555.2	6	5862.9	30
5472.3	1000	6991.9	150	2552.8	5	5837.3	400
5452.9	1000	6985.8	200	2551.2	2	5230.2	40
5451.5	1000	6916.5	200	2534.8	20	4811.6	50
5357.6	1000	6857.1	200	Spark:		4792.6	200
5271.9	2000	6846.6	500	4172.0	1000	4488.2	40
5266.4	1000	6752.6	300	2830.1	2	4437.2	50
5215.0	1000	4894.3	200	2418.6	5	4315.0	40
4435.5	2000	4873.3	200	2371.3	5	4241.7	40
				2338.2	4		

Gold (cont.)		Holnium Ho		Iridium Ir			
4065.0	50	5982.9	200	2389.5	50	5283.6	400
3897.8	30	5973.5	125	2340.1	10	5269.5	800
3553.5	20	5955.9	100	2324.9	3	5266.5	500
3320.1	20	5948.0	200	2324.4	5	5232.9	800
3230.6	15	5933.7	200	2306.8	25	5167.4	700
3204.7	50	5921.7	200	2283.7	2	4925.2	1000
3194.7	25	5882.9	200	2241.1	10	4920.5	500
3122.7	500	5860.2	200	2230.6	5	4903.3	500
3029.2	25	5691.4	200	2229.8	2	4528.6	600
2905.9	10	5674.7	200	2211.1	3	4476.0	500
2748.2	40	5566.5	100			4466.5	500
2675.9	250	4350.7	40	Iridium Ir		4459.1	400
2427.9	400	4254.4	100			4442.3	400
		4163.0	100	4728.8	150	4427.3	500
		4152.5	30	4616.3	200	4422.5	300
Hafnium Hf		4127.1	150	4568.0	100	4415.1	600
		4120.2	50	4548.4	100	4404.7	1000
6980.9	100	4108.6	100	4545.6	200	4383.5	1000
6818.9	100	4103.8	400	4532.8	80	4375.9	500
6644.6	100	4053.9	400	4496.0	40	4325.7	1000
5311.6	100	4045.4	200	4495.3	100	4315.0	500
5298.0	80	4040.8	150	4492.1	35	4307.9	1000
5264.9	50	3891.0	200	4478.4	200	4299.2	500
5040.8	100	3854.0	10	4450.1	60	4294.1	700
4664.1	50			4426.2	400	4282.4	600
4655.1	50			4403.7	300	4271.7	1000
4544.0	20	Indium In		4399.4	400	4271.1	400
4541.2	15			4392.5	100	4260.4	400
4540.9	50	6847.7	60	4377.0	100	4250.7	400
4518.2	10	5728.2	50	4311.4	300	4202.0	400
4499.6	15	4511.3	5000	4310.5	150	4143.8	400
4486.1	25	4101.7	2000	4301.6	200	4063.5	400
4461.1	25	3258.5	500	4286.6	200	4045.8	400
4457.3	25	3256.0	1500	4268.1	200	3969.2	600
4443.0	15	3039.3	1000	4259.1	200	3930.2	600
4438.0	30	2932.6	500	4172.5	150	3927.9	500
4370.9	30	2753.8	300	4166.0	150	3922.9	600
4350.5	20	2710.2	800			3920.2	500
4336.6	30	2560.2	150	Iron Fe		3902.9	500
4294.7	25	2521.3	30			3899.7	500
4093.1	25	2468.0	25	5429.6	500	3895.6	400
		2460.0	10	5415.2	500	3859.9	1000
				5371.4	700		

Iron
(*cont.*)

3749.4	1000
3734.8	1000
3719.9	1000
3681.1	1000
3490.5	400

Lanthanum
La

7066.2	400
5602.5	300
5301.9	300
5211.8	300
5183.4	400
4999.4	400
4921.7	500
4920.9	500
4899.9	400
4743.0	300
4728.4	400
4574.8	300
4333.7	800
4286.9	400
4238.3	500
4123.2	500
4086.7	500
4077.3	600
4042.9	400
4021.6	400
3995.7	600
3988.5	1000
3949.1	1000
3929.2	400
3921.5	400

Lead
Pb

9674.5	100
9604.0	50
8272.8	200
7229.1	50
6001.8	40
5895.7	20
5201.4	10
5005.4	20
4168.0	20
4062.1	20
4057.8	2000
4019.6	6
3739.9	150
3683.4	300
3639.5	300
3572.7	200
2873.3	100
2833.0	500
2802.0	250
2663.1	300
2393.7	2500
2169.9	1000
2159.5	15
2111.7	20
2088.4	30

Lithium
Li

8126.5	1000
6707.8	3000
6240.1	300
6103.6	2000
4971.9	500
4636.1	3
4602.8	800
4273.2	200
4132.2	400
3985.7	100
3915.0	20
3794.7	60
3718.7	30
3232.6	1000
2741.3	200
2562.5	150
2475.2	100
2425.6	30
2394.4	5

Lutecium
Lu

9914.9	100
8610.9	125
8508.0	100
8459.1	150
7125.8	125
6611.7	100
6463.1	400
6221.8	500
6055.0	150
6004.5	400
5736.5	150
5476.6	500
5421.9	50
5402.5	150
5125.0	200
5001.1	100
4994.1	250
4518.5	300
4430.4	30
4281.0	40
4124.7	200
4054.4	25
2603.2	5
2549.5	20
2536.9	10

Magnesium
Mg

8806.7	100
7779.9	5
5528.4	60
5183.6	500
5172.6	200
5167.3	100
4703.0	8
4433.9	8
3838.2	300
3832.3	250
3829.3	100
3096.8	150
3092.9	125
3091.0	80
2915.5	20
2852.1	300
2851.6	25
2802.6	150
2795.5	150
2779.8	40
2778.2	25
2776.6	30
2736.5	30
2733.5	25
2672.5	20

Manganese
Mn

8740.9	500
8703.7	200
7326.5	500
7283.8	400
6021.7	80
6016.6	80
6013.4	100
5341.0	200
4823.5	400
4783.4	400
4762.3	100
4754.0	400
4739.1	150
4727.4	150
4709.7	150
4701.1	100
4605.3	150
4498.8	150
4414.8	150
4034.4	250
4033.0	400
4030.7	500
2801.0	600
2798.2	800
2794.8	1000

Mercury
Hg

5790.6	500
5769.5	600
5460.7	500
4358.3	2000
4347.4	200
4339.2	150
4077.8	150
4046.5	200
3906.4	25
3663.2	500
3662.8	50
3650.1	200
3341.4	100
3131.8	200
3131.5	400
3125.6	200
3021.4	80
2967.2	100
2893.5	40
2752.8	40
2698.8	25
2652.0	100
2536.5	2000
2534.7	30
2482.7	25

Molybdenum
Mo

6733.9	100
6707.8	300
6619.1	300
6424.3	100
6030.6	300
5928.8	100
5888.3	150
5858.2	200
5751.4	125
5632.4	100
5570.4	200
5533.0	200
5506.4	200
4760.1	125
4731.4	100
4707.2	125
4381.6	150
4293.2	125
4284.5	125
3902.9	1000
3864.1	1000
3798.2	1000
3193.9	1000
3170.3	1000
3132.5	1000

Neodymium
Nd

6385.1	100
5804.0	100
5688.5	150
5669.7	40
5620.5	200
5594.4	150
4924.5	80
4896.9	60
4859.0	60
4646.6	60
4641.1	80
4516.3	70
4465.6	20
4462.9	60
4458.5	25
4451.9	50
4451.5	100
4444.2	20
4374.9	20
4325.7	100
4303.5	100
4256.4	40
4040.7	40
4012.2	80
3590.3	400

Nickel
Ni

7422.3	600
7122.2	1000
6256.3	600
6191.1	500
6176.8	400
6175.4	300
5476.9	400
5084.0	300
5081.1	150
5080.5	200
5035.3	300
4984.1	500
4980.1	500
4937.3	400
4714.4	1000
4401.5	1000
3973.5	800
3858.3	800
3619.3	2000
3597.7	1000
3566.3	2000
3515.0	1000
3492.9	1000
3446.2	1000
2414.7	1000

Osmium
Os

5523.5	100
4793.9	300
4663.8	100
4616.7	150
4551.2	150
4550.4	150
4548.6	100
4484.7	100
4447.3	200
4420.4	400
4394.8	150
4311.3	150
4260.8	200
4211.8	150
4202.0	100
4175.6	100
4135.7	200
4112.0	150
4091.8	100
4088.4	100
3977.2	300
3963.6	500
3782.1	400
3752.5	400
2909.0	500

Palladium
Pd

5695.0	50
5670.0	100
5619.4	50
5542.7	100
5395.2	50
5295.6	200
5163.8	300
5110.8	100
4788.1	200
4541.1	15
4473.5	60
4212.9	500
4169.8	200
4140.8	100
4087.3	500
3958.6	500
3634.6	2000
3609.5	1000
3516.9	1000
3481.1	500
3441.3	800
3433.4	1000
3421.2	2000
3302.1	1000
3242.7	2000

Phosphorus
P

9896.7	100

Characteristic Lines of the Elements

Palladium
(cont.)
9750.7 70
9593.5 70
9525.7 100
4246.8 70
4222.1 300
2554.9 60
2553.2 80
2535.6 100
2534.0 50
2154.0 15
2152.9 12
2149.1 15
2136.1 15
2033.4 15
2032.4 12
2024.5 12
2023.4 15
Tube:
5425.9 150
5253.4 300
4601.9 300
4589.9 300
3490.4 70
3488.7 70

Platinum
Pt
6760.0 100
5478.4 50
5475.7 60
5368.9 50
5301.0 150
4552.4 60
4520.9 40
4498.7 100
4445.5 20
4442.5 800
4437.2 25
4327.0 80
4192.4 100
4164.5 100

4118.6 400
3922.9 100
3638.7 250
3626.1 300
3301.8 300
3251.9 100
3139.3 300
3064.7 2000
2997.9 1000
2733.9 1000
2659.4 2000

Potassium
K
9591.8 50
7698.5 5000
7664.9 9000
6938.9 500
6911.3 300
5832.0 50
5812.5 30
5801.9 50
5782.6 60
5359.5 40
5342.9 30
5339.6 40
5323.2 40
5112.2 30
5099.1 25
5097.1 25
5084.2 20
4965.0 15
4047.2 400
4044.1 800
3447.7 100
3446.7 150
3217.5 50
3217.0 100
3102.0 50

Praseodymium
Pr
6055.1 100

5707.6 100
5259.7 125
5173.8 100
5129.5 100
4951.3 150
4848.5 125
4822.9 125
4744.9 100
4736.6 125
4684.9 125
4651.5 125
4628.7 200
4534.1 150
4510.1 200
4496.4 200
4429.2 200
4179.4 200
4164.1 200
4143.1 200
4118.4 250
4100.7 200
4062.8 150
3994.8 300
3989.7 200

Rhenium
Re
7912.9 400
7640.9 400
6829.9 200
6813.4 200
5834.3 200
5776.8 300
5752.9 200
5667.9 100
5563.2 150
5532.6 100
5377.0 300
5278.2 100
5275.5 500
5270.9 200
4923.9 150
4889.1 2000

4791.4 200
4513.3 300
4394.3 100
4392.4 100
4227.4 200
4136.4 150
4133.4 200
3460.4 1000
3426.1 30

Rhodium
Rh
6965.6 200
6752.3 150
5983.5 200
5806.9 100
5686.3 100
5599.4 300
5424.0 100
5390.4 125
5379.0 100
5354.3 300
5193.1 200
5175.9 200
5155.5 150
5090.6 150
4528.7 500
4374.8 1000
4288.7 400
4135.2 300
4128.8 300
3657.9 500
3570.1 400
3528.0 1000
3498.7 500
3478.9 500
2434.8 1000

Rubidium
Rb
8271.7 100
8271.4 200
7947.6 5000

Rhodium (cont.)		Samarium Sm		Silicon Si		Silver Ag	
7925.5	70	4554.5	1000	5711.7	100	2524.1	400
7925.2	100	4460.0	150	5700.2	400	2519.2	300
7800.2	9000	4410.0	150	5686.8	200	2516.1	500
7759.4	400	4397.7	150	5671.8	300	2514.3	300
7757.6	1000	4390.4	150	5526.8	100	2506.8	300
7618.9	1000	4199.9	150	5520.4	80	2124.1	200
7408.1	500	4144.1	150	5099.2	100	2122.9	10
7279.9	400	3498.9	500	5085.5	80		
6299.2	300	3436.7	300	5083.7	100	5471.5	500
6298.3	1000			5081.5	100	5465.4	1000
6206.3	800	8485.9	400	4743.8	100	5209.0	1500
6159.6	400	8305.7	500	4741.0	100	4668.4	200
6070.7	600	8068.4	800	4737.6	100	4476.0	40
5724.4	600	8026.3	500	4734.0	100	4396.3	100
5653.7	200	8025.1	400	4729.2	100	4212.6	150
5578.7	150	7928.1	800	4670.4	100	4210.9	200
5431.5	100	6955.2	400	4415.5	100	4055.2	800
4215.5	1000	6861.0	500	4400.3	150	3981.6	30
4201.8	2000	6860.9	800	4374.4	100	3382.8	1000
3591.5	80	6856.0	300	4023.6	100	3280.6	2000
3587.0	200	6790.0	300	4023.2	100	2824.3	150
3350.8	150	6734.8	400	4020.3	50	2660.4	30
		6734.0	400			2506.6	15
Ruthenium Ru		6731.8	500	Silicon Si		2447.9	30
		6604.5	200			2437.7	60
6982.0	200	6589.7	400	9413.5	100	2413.1	50
6923.2	300	6570.6	200	8752.1	200	2375.0	300
6911.4	100	6569.3	500	8742.6	100	2331.3	18
6824.0	200	6542.7	200	8556.6	100	2324.6	15
6690.0	300	5600.8	200	7943.9	500	2320.2	15
5309.2	125	4745.6	250	7932.2	300	2317.0	15
5171.0	150	4424.3	300	7918.3	200	2312.4	25
5155.1	125	4329.0	300	7681.3	100	2309.6	150
5126.5	125	4318.9	300	7423.5	500		
4757.8	125	4251.7	200	7416.0	200	Sodium Na	
4709.4	150			7409.1	100		
4684.0	100	Scandium Sc		7405.5	300		
4681.7	100			7289.2	200	8194.8	1000
4647.6	125	7741.1	50	7165.6	100	8183.2	500
4645.0	100	6835.0	25	5948.5	50	6160.7	500
4635.6	125	6819.5	20	5754.2	40	6154.2	500
				2881.5	500	5895.9	5000
				2528.5	400		

Characteristic Lines of the Elements

Sodium	4337.8 150	2142.7 600	4005.5 100
(cont.)	4305.4 40	2081.0 400	3976.8 150
5889.9 9000	4215.5 300	2039.7 300	3958.3 100
5688.2 300	4161.7 30	2001.9 600	3925.4 150
5682.6 80	4077.7 400	Tube:	3899.1 200
5675.7 150	4030.3 40	6422.9 70	
5670.1 100	3351.2 300	6709.0 50	**Thallium**
5153.6 600		5708.0 250	**Tl**
5149.0 400	**Tantalum**	5649.3 250	6713.6 100
4982.8 200	**Ta**	2548.2 5	6552.6 200
4978.5 15	6928.5 150	2544.8 5	6549.7 300
4751.8 20	6927.3 150	2543.7 10	5949.0 20
4748.0 15	6774.2 100	2537.8 300	5350.4 5000
4668.5 200	6771.7 100	2537.1 5	3775.7 3000
4664.8 80	6675.5 400	2533.4 5	3529.4 1000
4498.7 70	6673.7 200	2530.7 30	3519.2 2000
4494.2 60	6516.0 200	2530.0 15	3229.7 2000
4393.4 20	6514.3 200	2527.1 10	2921.5 200
4390.1 15	6485.3 500	2525.7 5	2918.3 400
4344.5 3	6256.5 300	2524.9 15	2826.1 200
3302.3 600	6154.5 200		2767.8 400
2680.4 40	6020.7 300	**Terbium**	2710.6 30
	4691.9 400	**Tb**	2709.2 400
Strontium	4669.1 300	5685.7 40	2609.7 30
Sr	4661.1 300	4563.6 50	2608.9 80
7673.0 200	4619.5 300	4511.5 40	2585.5 30
7621.5 100	4574.3 300	4509.0 30	2580.8 30
7309.4 200	4551.9 400	4493.0 100	2552.9 10
7167.2 100	4530.8 300	4356.8 60	2552.5 80
7070.1 1000	4511.5 300	4342.5 75	2517.4 30
6892.5 100	3318.8 125	4338.4 100	2379.6 100
6878.3 500	3311.1 300	4336.5 40	2315.9 60
6791.0 200	2902.0 1000	4332.1 40	2237.8 60
6643.5 100	2806.5 200	4326.4 150	
6617.2 150	2806.3 300	4278.5 200	**Thorium**
6550.2 100		4276.7 50	**Th**
5543.3 30	**Tellurium**	4258.2 30	5028.6 40
5540.0 20	**Te**	4226.4 50	4544.5 8
5534.8 20	3175.1 30	4200.9 40	4541.6 8
5521.8 50	2385.7 600	4100.9 50	4541.2 6
5480.8 100	2383.2 500	4052.8 40	4540.4 8
4876.3 200	2208.8 200	4032.2 30	4510.5 30
4607.3 1000	2159.7 100	4031.6 50	4499.9 12

Thorium (cont.)

4487.5	20
4486.6	10
4485.7	10
4480.8	25
4472.3	10
4465.3	30
4461.0	10
4454.5	8
4447.8	12
4443.0	10
4440.8	20
4440.5	15
4439.1	20
4320.5	8
4320.1	12
4281.0	20
4277.3	20
4005.5	20

Thulium Tm

8017.9	200
7558.3	200
7490.2	100
6845.7	200
6844.2	250
6779.7	300
6604.9	300
6460.2	400
5675.8	100
5631.4	80
5307.1	100
5034.2	100
4615.9	200
4529.3	80
4522.5	200
4481.2	400
4386.4	200
4359.9	300
4242.1	500
4203.7	250
4199.9	100
4187.6	300
4105.8	300
4094.1	300
3996.5	200

Tin Sn

9850.5	500
9805.3	300
9616.4	150
8708.3	20
8552.6	500
8114.1	200
7754.9	100
5631.6	50
4524.7	500
4511.3	200
3262.3	400
3175.0	500
3034.1	200
2913.5	100
2863.3	300
2839.9	300
2706.5	200
2495.7	100
2483.4	125
2429.4	200
2421.6	150
2354.8	150
2317.2	100
2268.9	100
2246.0	100

Titanium Ti

8468.5	300
8467.1	300
6261.0	300
6258.7	300
6258.1	200
6146.2	400
5064.6	150
5039.9	125
5038.4	100
5036.4	125
5035.9	125
5025.5	100
5024.8	100
5022.8	100
5020.0	100
5016.1	100
5014.2	100
5007.2	200
4999.5	200
4991.0	200
4981.7	300
4305.9	300
3729.8	500
3653.4	500
3642.6	300

Tungsten W

5735.0	50
5514.6	50
5492.3	50
5224.6	50
5204.5	40
5203.2	30
5192.7	30
5183.9	20
5053.2	60
4757.5	60
4680.5	150
4659.8	200
4551.8	35
4513.3	30
4441.8	20
4348.1	50
4302.1	60
4294.6	50
4074.3	50
4008.7	45
3617.5	35
3170.2	15
2947.3	12
2946.9	20
2397.0	18

Uranium U

6449.1	100
6395.4	100
5915.3	125
5511.5	30
5492.9	60
5481.2	30
5475.7	20
5027.3	40
4543.6	50
4512.3	20
4477.7	20
4472.3	50
4465.1	20
4440.7	20
4372.7	18
4372.5	15
4341.6	50
4306.7	40
4282.0	30
4246.2	30
4241.6	40
4211.6	30
4171.5	30
4124.7	30
4042.7	40

Vanadium V

6199.1	100
6081.4	100
6077.3	300
6039.7	100
5978.9	100
5924.5	250
5846.2	100
5830.7	100

Characteristic Lines of the Elements 185

Vanadium (cont.)					
5817.5	100	5352.9	100	4083.7	50
5737.0	100	5277.0	**200**	3982.5	60
5731.2	250	5074.3	200	3950.3	60
5727.0	150	4935.5	200	3710.2	80
5706.9	200	4576.2	90	3300.7	100
5703.5	200	4515.1	45		
5698.5	300	4439.2	45		
5670.8	150	4277.7	25		
5627.6	200	4089.6	50		
5192.9	100	4077.2	30		
4384.7	125	3987.9	1000		
4392.2	200	3694.2	500		
3855.8	200	3289.8	1000		
3676.6	300	3289.3	500		
3185.3	500				
3183.9	500				
3183.4	200				

Bands:
6132.1 200
6003.6 200
5987.6 300
5973.0 600

Zinc
Zn
6362.3 1000
5181.9 200
4924.0 15
4810.5 400
4722.1 400
4680.1 300
4629.8 35
4057.7 80
3345.5 500
3345.0 800
3302.9 700
3302.5 800
3282.3 500
3072.0 200
3035.7 200
2770.9 300
2770.8 300
2756.4 200
2712.4 300
2608.6 300
2557.9 10
2502.0 20
2161.9 100
2138.5 800
2025.5 200

Zirconium
Zr
7169.0 150
7097.7 100
6313.0 200
6143.2 300
6134.5 300
6127.4 500
4772.3 100
4739.4 100
4687.8 125
4302.8 100
4241.6 100
4241.2 100
4240.3 100
4229.3 100
4227.7 150
4149.2 100
4081.2 150
4072.7 100
4064.1 100
4055.0 100
3958.2 500
3929.5 100
3636.4 200
3601.1 400
2571.3 300

Ytterbium
Yb
7699.4 2000
7527.5 80
6799.6 1000
6667.8 1000
6489.1 800
6417.9 125
6400.4 200
6274.7 100
5720.0 300
5556.4 1500
5539.0 200

Yttrium
Yt
9231.5 80
6435.0 150
5630.1 80
5581.8 100
5527.5 100
5466.4 150
4527.2 40
4505.9 50
4398.0 150
4374.9 150
4358.7 60
4348.7 100
4174.1 100
4142.8 100
4128.3 150
4102.3 150

Chapter VII

WAVE-LENGTH TABLE-CHART

In the following Table-Chart the distance between printed lines represents the space of one angstrom. Thus the spacing of spectral lines may be seen at once, as from a chart. If one wave-length number, for example, is five printed lines below another, the spectral lines which the numbers represent are five angstroms apart, if ten lines below, they are ten angstroms apart, and so on.

The strongest spectral lines or bands for each angstrom space are given with the decimal indicating still more exactly the position of the line which has the greatest intensity. The spectral lines customarily seen in common rocks and ores are placed in the column to the left of the wave-length numbers for use as spotters.

The visible range from 3900 to 7100 Å is covered; only arc lines are included. The numbers in parentheses indicate intensities on a scale of 1-10000. At the red end of the table a printed line will sometimes give all of the known spectral lines for that angstrom space. At the violet end, where the spectral lines are crowded, only the few most important of perhaps 30 or 40 known lines are given. All lines of intensity 100 or more have been included, and many weaker ones.

Since each printed line represents an angstrom space, the peculiar arrangement of spectral lines in groups may be read without computation. Sometimes the most intense lines of an element, because of the interfering lines of other elements, are not the best for identification. In such cases, weaker lines in the clear, or significant groups of weaker lines are more practical for rapid analysis.

Do not depend upon single lines; always check neighboring lines of the same or of other elements.

6900

99.9 Fe (25)	Yb (15)	38.9 K (500)	Sm (10)
98.8 Ir (15)	Dy (3)	37.8 Co (150)	Gd (8)
97.2 Co (200)	Hf (2)	36.0 Ta (2)	
96.7 Gd (200)	Ti (15)	35.8 Ti (8)	Cu (8)
95.3 Ta (200)	Sb (5)	34.9 La (50)	Mo (15)
94.3 Zr (15)	Er (6)	33.1 Ti (12)	Yt (7)
93.1 Gd (20)	Sm (15)	32.3 Zr (12)	Gd (5)
92.8 Sm (2)		31.1 Mn (20)	Mo (15)
91.9 Gd (150)	Bi (10)	30.4 Sm (50)	
90.3 Cb (100)	Zr (50)	29.8 Ir (50)	Sm (40)
89.8 Mn (80)	Th (20)	28.5 Ta (150)	Zn (15)
88.4 Sm (40)	Mo (30)	27.3 Ta (150)	Sm (40)
87.3 Sm (10)	Sm (4)	26.1 Ti (35)	Gd (10)
86.0 Cb (20)	Ce (15)	25.2 La (100)	Cr (50)
85.8 Gd (200)	Re (20)	24.1 Cr (60)	Ce (20)
84.9 Os (10)	Sm (10)	23.2 Ru (300)	Nd (4)
83.4 Cs (25)	Ta (20)	22.2 Zr (8)	Co (5)
82.0 Ru (200)	Yb (3)	21.0	
81.0 Be (15)	Cr (8)	20.0 Cu (100)	Gd (30)
80.9 Hf (100)	Gd (25)	19.0 Sm (30)	Gd (5)
79.1 Rh (25)	Cr (20)	18.3 Cb (80)	La (20)
(60) Fe 78.4 Cr (125)	Mo (25)	17.3 Lu (50)	La (25)
77.5 Sm (10)	Fe (4)	16.5 Gd (200)	Fe (35)
76.5 Si (25)	Gd (20)	15.3 U (3)	
75.0 Cb (12)	Sm (10)	14.5 Ni (300)	Mo (40)
74.5 V (2)		13.1 Ti (7)	Sm (3)
73.2 Cs (500)	Eu (8)	12.7 Sm (3)	Dy (2)
72.4 Cb (20)	Rh (6)	11.3 K (300)	Ru (100)
71.5 Re (150)	Gd (20)	10.8 Co (15)	Eu (10)
70.4 Hf (5)		09.8 Sm (50)	
69.6 Sm (15)	Gd (8)	08.0 Cb (40)	Co (30)
68.6 Sm (25)	Mn (5)	07.3 Zr (12)	Sm (2)
67.6 Re (6)		06.2 Sm (20)	Nd (10)
66.1 Ta (150)	Zr (40)	05.9 Cu (40)	Ru (25)
65.6 Rh (200)	Sm (8)	04.5 Sm (40)	Zr (10)
64.6 Nd (5)	Gd (5)	03.7 Eu (200)	U (2)
63.9 Ti (5)	Yb (2)	02.1 Ta (150)	Cb (60)
62.0		01.5 Co (5)	Os (3)
61.4 Mo (10)	Ba (2)	00.5 Ta (80)	Gd (60)

6800

60.6 Mo (12)	Ce (10)	99.0 Mo (12)	Tb (6)
59.2 Gd (20)	Ir (10)	98.2 Eu (30)	Mo (15)
58.0 La (25)	Ti (10)	97.7 Ta (30)	Er (6)
57.7 Gd (20)	Mo (5)	96.3 Tb (10)	Yt (5)
56.2 Ca (20)	Band	95.7 Er (6)	
55.2 Sm (400)	Ni (80)	94.5 Ce (6)	V (2)
54.5 La (15)	Gd (3)	93.3 Ir (30)	Ce (5)
53.8 Zr (80)	Ta (50)	92.5 Sr (100)	Ta (30)
52.4 La (6)	Sm (3)	91.4 Sm (40)	
51.2 Ta (100)	Fe (10)	90.9 Cu (8)	Sc (3)
50.5 Sm (100)	Yt (20)	89.9 Cu (8)	Sc (7)
49.2 Sm (15)		88.2 Zr (25)	Ir (25)
48.1 Sm (20)	Zr (15)	87.6 Gd (35)	Mn (30)
47.3 Mo (15)	Gd (6)	86.3 Cb (30)	Mo (25)
46.8 Ta (20)	Cb (15)	85.1 Sm (30)	Fe (10)
(60) Fe 45.2 Gd (20)		84.9 Sm (6)	Er (2)
44.9 Er (6)	Gd (5)	83.0 Cr (70)	Zr (8)
43.7 Ti (15)	Th (10)	82.3 Cr (20)	Sm (2)
42.5 Mn (100)	U (2)	81.6 Cr (15)	Cu (8)
41.5 Sm (50)	Nd (2)	80.0 Er (6)	Tb (4)
40.9 Cb (10)	Nd (2)	79.9 Rh (30)	Cb (15)
39.5 Ce (8)	Ta (4)	78.3 Sr (500)	Gd (6)

	77.4 Ta (30)	Sm (25)
	76.3 Cb (80)	Ni (25)
	75.2 Ta (200)	Sm (20)
	74.1 Tb (6)	Hf (5)
	73.2 Sm (20)	Ti (10)
	72.4 Co (200)	Sm (100)
	71.5 Yb (10)	V (2)
	70.4 Cs (200)	Cb (20)
	69.0	
	68.5 Th (2)	Sm (2)
(100)Ba	67.8 Sm (15)	
	66.2 Ta (200)	Sm (8)
(200)Ba	65.7 Ta (5)	Er (4)
	64.5 Eu (1000)	
	63.6 Mo(3)	Mn(2)
	62.8 Sm (100)	Fe (7)
(50)Ti	61.0 Sm (500)	Ni (5)
	60.9 Sm (800)	Ti (20)
	59.0 La (4)	U (2)
	58.3 Co (25)	Fe (15)
	57.1 Gd (200)	Nd (8)
	56.0 Sm (300)	Ce (4)
(60)Fe	55.1 Ti (20)	Hf (7)
	54.5 Sm (60)	Tm(20)
	53.5 Sm (10)	Zr (6)
	52.5 Zr (6)	Gd (5)
	51.0	
	50.0 Hf (20)	Ta (5)
	49.3 Cb (12)	Gd (5)
	48.8 Sm (40)	Mo(15)
	47.7 In (60)	Eu (10)
	46.6 Gd (500)	Sm (100)
	45.7 Tm(200)	Yt (15)
	44.2 Tm(250)	Sm (150)
	43.6 Fe (30)	Rh (5)
	42.0 Ni (60)	Pt (10)
(50)Fe	41.3 Eu (50)	Sm (25)
	40.9 Cu (3)	
	39.6 Sm (40)	Sm (15)
	38.8 Mo(30)	Co (15)
	37.2 Sm (20)	La (15)
	36.6 Gd (3)	U (2)
	35.0 Sc (25)	Dy (8)
	34.4 Eu (30)	La (25)
	33.9 Mn(40)	Pd (10)
	32.8 Zr (10)	V (5)
	31.1 Tm(20)	Ru (6)
	30.0 Ir (50)	Sm (15)
	29.9 Re (200)	Sm (100)
	28.1 Cb (150)	Gd (150)
	27.8 Sm (15)	Rh (15)
	26.9 U (25)	Hf(10)
	25.2 Cs (15)	Er (6)
	24.0 Ru (200)	Ta (5)
	23.7 La (50)	Sm (15)
	22.7 Gd (15)	Eu (15)
	21.9 Sm (10)	La (5)
	20.9 Gd (50)	Sm (15)
	19.5 Co (20)	Sc (20)
	18.9 Hf (100)	Ce (5)
	17.0 Sc (10)	
	16.1 Eu (150)	Gd (40)

15.5 Sm (10)	Yt (3)
14.9 Co (150)	Gd (30)
13.4 Re (200)	Ta (200)
12.4 V (8)	Zr (6)
11.3 Zr (4)	Re (3)
10.4 Ta (40)	Fe (15)
09.2 Sm (2)	
08.9 Co (25)	Sm (20)
07.8 Ce (10)	Sm (5)
06.8 Fe (8)	Os (8)
05.5 Ru (5)	Re (3)
04.0 Nd (15)	Sm (10)
03.1 Sm (30)	Yt (4)
02.7 Eu (500)	Sm (25)
01.7 Sm (15)	Re (5)
00.7 Gd (3)	Yb (3)

6700

99.6 Yb (1000)	
98.6 Pr (20)	Ca (6)
97.0	
96.8 Sm (10)	Er (6)
95.4 Yt (12)	Cb (10)
94.2 Sm (200)	Tb (10)
93.7 Yt (70)	Lu (40)
92.5 Sm (30)	Tb (4)
91.0 Sr (200)	Os (10)
90.0 Sm (300)	U (10)
89.2 Hf (50)	Co (15)
88.9 Ta (50)	Tm(20)
87.5 Eu (30)	Ru (20)
86.3 Gd (125)	Fe (7)
85.0 V (15)	Tb (8)
84.8 Co (25)	Pd (12)
83.3 Gd (40)	
82.5 Eu (100)	Sm (30)
81.1 Sm (50)	Au (3)
80.0 Sm (40)	Ce (5)
79.7 Tm(300)	Sm (15)
78.6 Sm (200)	Cd (30)
77.7 Gd (3)	Yb (2)
76.8 U (5)	Er (4)
75.0 Ru (40)	Sm (15)
74.2 Ta (100)	La (100)
73.0 Hf (5)	Er (4)
72.3 Ni (200)	Gd (20)
71.0 Co (200)	Ta (100)
70.3 Sm (10)	Ta (3)
69.1 Zr (50)	Hf (10)
68.7 Yb (80)	Er (4)
67.7 Ni (300)	Co (35)
66.5 Sm (50)	Ru (30)
65.9 Dy (6)	
64.4 Ce (2)	
63.5 Mo(8)	Nd (6)
62.4 Cr (50)	Zr (50)
61.1 Re (10)	Gd (5)
60.0 Pt (100)	V (2)
59.2 Sm (15)	Er (8)
58.1 Co (25)	Eu (4)
57.7 Cr (8)	
56.9 Sm (50)	Co (25)
55.8 Ta (10)	Tb (4)

	54.6 Hf (60)	Sm (50)	(600)Ba	93.8 Eu (500)	Sm (100)
	53.9 Gd (100)	La (100)		92.8 La (50)	Eu (40)
	52.6 Gd (300)	Rh (150)		91.5 Mn(25)	Mo(5)
	51.2 Re (50)	Sm (2)		90.0 Ru (300)	Mo(20)
(50)Fe	50.1 Eu (10)	Sm (2)		89.5 Tb (4)	Ce (3)
	49.6 Nd (10)	Cu (8)		88.1 Zr (10)	Sm (4)
	48.1 La (25)	Ir (4)		87.7 Sm (80)	Yt (50)
	47.2 Pr (20)	Dy (4)		86.0 Ir (20)	Ce (8)
	46.2 Mo(50)	Mo(20)		85.2 Eu (80)	Pd (2)
	45.1 Nd (5)	Yb (2)		84.0 Ta (40)	Co (60)
	44.9 Eu (100)	Cr (6)		83.3 Re (15)	U (6)
(100)Ti	43.1 Sm (2)			82.0 Eu (8)	Sm (5)
	42.5 Nd (10)	Co (5)		81.2 Gd (100)	Sm (60)
	41.4 Sm (40)	Cu (50)		80.1 Nd (2)	
	40.7 Ta (80)	Nd (10)		79.1 Sm (160)	Gd (25)
	39.8 Cb (80)	Sc (3)		78.8 Co (125)	Yb (20)
	38.9 Sb (2)		(250)Fe	77.9 Cb (200)	Lu (40)
	37.7 Nd (10)	Sc (5)		76.9 U (8)	
	36.0 Yt (12)	Pr (15)	(500)Ba	75.2 Ta (400)	Ce (3)
	35.3 Sm (100)	Lu (5)		74.6 Sm (2)	
	34.8 Sm (800)	Cr (12)		73.7 Ta (200)	Pr (40)
	33.9 Mo(100)	Fe (5)		72.2 Cu (15)	Ru (5)
	32.0 Nd (10)	Eu (4)		71.4 Sm (50)	La (10)
	31.8 Sm (500)	Ru (4)		70.7 Sm (3)	Nd (2)
	30.7 Gd (100)	Ru (25)		69.2 Cr (80)	Sm (10)
	29.5 Os (30)	Ce (10)		68.7 Lu (4)	Ir (4)
	28.9 Nd (81)	Ce (8)		67.8 Yb (1000)	
	27.8 Gd (50)	Yb (30)		66.5 Ti (30)	Cr (8)
	26.7 Fe (15)	Sm (15)		65.2 Re (25)	Ce (10)
	25.9 Sm (30)			64.6 Nd (8)	Yt (6)
	24.7 Sm (30)		(70)Fe	63.1 Ru (100)	Co (3)
	23.2 Cs (500)	Cb (100)		62.2 Th (10)	Ta (10)
	22.7 Er (6)	Ir (5)		61.0 Cr (100)	La (70)
	21.3 Tm(60)	Sm (10)		60.8 Cb (300)	
	20.9 Sm (5)	Co (4)		59.6 Mo(20)	Ir (5)
	19.5 Sm (5)	Hf (2)		58.6 Tm(30)	Dy (4)
	18.1 Gd (40)	Ru (15)		57.7 Tm(70)	
	17.6 Ca (800)	Zr (12)		56.1 Sm (100)	Pr (30)
	16.6 Ti (15)	Hf (1)		55.6 F (100)	Band
	15.3 Cr (8)	Fe (5)	(50)Ba	54.0 Sm (4)	Dy (3)
	14.6 Sc (5)	Ta (4)		53.0	
	13.6 Tl (100)	Yt (15)		52.3 Re (80)	F (100)Band
	12.6 Sm (50)	Sm (25)		51.6 Sm (80)	Gd (15)
	11.2 Re (10)	Sm (5)		50.8 La (100)	Mo (80)
	10.4 Pt (50)	Eu (25)		49.0 Sm (50)	Mo(7)
	09.4 La (150)	Cb (15)		48.7 F (100)	Band
	08.1 W (20)	V (2)		47.0 Hf (30)	Ni (5)
	07.8 Li (3000)Mo(300)	Co (200)		46.8 Gd (10)	Sm (10)
	06.8 Sm (5)	Ta (5)		45.1 Eu (1000)	Gd (150)
	05.1 Fe (12)	Ta (3)		44.6 Hf (100)	La (40)
	04.3 Ce (25)	Gd (25)		43.6 Ni (300)	Sr (100)
	03.6 Sm (30)	Co (15)		42.0 F (50)	Band
	02.0 Gd (25)	Zr (6)		41.1 Ce (3)	Sm (3)
	01.2 Cb (100)	Eu (10)		40.0 Gd (40)	Sm (4)
	00.7 Ce (20)	Yt (12)		39.8 Sm (4)	Ta (2)
	6600			38.8 F (30)	Band
	99.2 Yt (18)	Yb (10)		37.1 Sm (60)	Nd (25)
	98.7 Al (10)	Sm (4)		36.5 Yt (8)	La (1)
	97.2 Sm (6)	Zr (4)		35.1 Co (25)	Sm (10)
	96.3 Al (10)	Cb (6)		34.3 Gd (150)	Cu (3)
	95.8 Eu (4)	Sm (3)	(60)Fe	33.7 Sm (10)	
	94.6 Sm (30)	Gd (20)	(150)Co	32.4 Sm (100)	F (300)Band

	31.2 La (3)			70.6 Sm (200)	Eu (15)
	30.6 Sm (50)	Cr (50)	(50) Fe	69.3 Sm (500)	Mo (10)
	29.4 F (200)	Band		68.4 Nd (5)	
	28.8 Sm (50)	Cs (35)		67.8 Eu (600)	Gd (25)
	27.8 Rh (30)	Tm (10)		66.7 Pr (20)	Tm (10)
	26.1 F (200)	Band	(50) Ti	65.6 Cu (15)	La (15)
	25.2 Sm (30)	Pd (4)		64.7 Gd (25)	Ta (20)
	24.7 Ir (15)	Mo (15)		63.4 Co (200)	Sm (20)
	23.7 Co (70)	Re (30)		62.9 Sm (50)	Hf (2)
	22.9 F (200)	Band		61.6 Ta (40)	Eu (6)
	21.3 Ta (200)	Cu (20)		60.4 Ru (9)	Nd (5)
	20.5 U (15)	Zr (6)		59.5 Ti (8)	Ce (3)
	19.1 Mo (300)	F (150) Band		58.1 Sb (12)	Sm (10)
	18.1 Ru (20)	Nd (8)		57.9 Hf (10)	Re (10)
	17.2 Sr (150)	Sm (50)	(150) Ti	56.0 Cb (6)	Er (6)
	16.5 La (125)	F (100) Band		55.6 Ce (15)	U (15)
	15.8 Nd (5)		(125) Ti	54.2 Th (5)	Re (3)
	14.8 Sm (50)	Cb (25)		53.0 Nd (8)	Pr (4)
	13.6 F (80)	Band		52.6 Tl (200)	Mo (5)
	12.1 Cr (12)	Lu (10)		51.4 Co (80)	Sm (25)
	11.9 Ta (300)	Lu (100)		50.2 Sr (100)	Ho (20)
	10.0 Gd (15)	Sb (4)		49.7 Tl (300)	Sm (100)
	09.1 Fe (25)	Nd (5)		48.2 Dy (3)	Hf (2)
	08.2 La (50)	Cr (25)		47.2 Gd (6)	Ta (5)
	07.0 Yb (20)	Cb (15)	(150) Fe	46.2 Ti (80)	Sr (25)
	06.1 Cb (15)	Ce (10)		45.4 Mg (10)	Re (6)
	05.1 Re (100)	Mn (30)		44.6 Cb (80)	Sm (80)
	04.9 Tm (300)	Sm (200)		43.1 La (125)	Sm (40)
	03.6 Eu (80)	Zr (20)		42.7 Sm (200)	U (8)
	02.6 Eu (2)			41.5 Er (6)	Hg (3)
	01.8 Sm (150)	Er (12)		40.4 Pr (15)	Ru (15)
	00.1 La (40)	Gd (8)		39.9 Nd (10)	
	6500			38.6 Yt (50)	Gd (50)
(100) Ti	99.1 Cu (25)	Ce (6)		37.9 Cr (35)	Mo (6)
	98.8 Zr (6)	Ni (5)		36.8 Sm (10)	Mo (7)
	97.5 Cr (40)	Sm (40)		35.3 Sm (5)	Yt (4)
	96.0 Sm (10)	Zr (4)		34.0 Mn (10)	Ce (5)
(1000) Ba	95.3 Co (150)	Nd (5)		33.4 Mn (20)	Sm (15)
	94.6 Cr (60)	Dy (3)		32.2 Sm (15)	Eu (4)
	93.8 Eu (400)	Fe (30)		31.4 V (25)	Th (15)
(150) Fe	92.9 Re (60)	Ni (5)		30.9 Gd (20)	Sm (10)
	91.6 Gd (40)	Sm (40)		29.7 Sm (50)	Cr (40)
	90.8 Mo (12)	U (5)		28.7 Ru (25)	Sm (5)
	89.7 Sm (400)	Nd (10)	(200) Ba	27.3 F (200)	Band
	88.9 Sm (100)	Th (10)		26.9 La (125)	Sm (50)
	87.5 Sm (30)	Hf (5)		25.3 Ce (5)	Sm (4)
	86.5 Cs (500)	Ni (40)		24.2 F (150)	Band
	85.2 Sm (150)	Nd (8)		23.4 Pt (80)	Lu (80)
	84.1 Sm (15)	Nd (5)		22.7 Eu (25)	Er (4)
	83.9 Th (10)	Er (10)		21.0	
	82.1 La (8)	U (4)		20.8 F (100)	Band
	81.8 Tb (6)	Sm (4)		19.6 Eu (500)	Rh (40)
	80.9 Cr (30)	Ba (20)		18.7 Mn (20)	U (15)
	79.3 Co (15)	Dy (8)		17.5 F (100)	Band
	78.5 La (100)	Pr (3)		16.0 Ta (200)	Sm (10)
	77.1 Re (50)	Sm (25)		15.2 Re (20)	
	76.8 Yt (10)	Zr (6)		14.3 Ta (200)	Nd (15)
	75.9 Sm (25)	Ti (20)		13.6 Ce (8)	
	74.8 Ta (200)	Sm (100)		12.0 F (300)	Band
	73.8 Gd (10)	Sm (10)		11.4 Re (60)	Th (4)
(50) Ca	72.7 Cr (25)	Nd (8)		10.4 Rh (25)	Sm (5)
	71.8 Nd (5)	Mo (5)		09.4 Sm (10)	Ce (8)

Wave-length Table-chart

(30)Ti	08.7 F (200)	Band			47.5 Sm (30)	Co (2)
	07.1 Sm (40)	Sm (15)			46.9 Ba (20)	Mo(20)
	06.3 Zr (25)	La (15)			45.7 Zr (30)	Ta (20)
	05.5 Ta (100)	F (100)Band			44.6 Ta (40)	Ru (25)
	04.1 V (25)	Co (15)			43.8 Ta (10)	Mn(10)
	03.9 Sr (35)	Zr (20)			42.5 U (2)	
	02.4 Ta (40)	Sm (40)			41.1 Lu (40)	Er (10)
	01.5 Eu (300)	Cr (35)			40.9 Mn(60)	Sm (8)
	00.7 Pr (10)	Nd (8)		(150)Ca	39.0 Co (80)	Eu (40)
	6400				38.4 Cd (2000)	W (3)
	99.6 Ca (30)	Co (2)			37.6 Eu (700)	Sm (10)
(60)Ba	98.6 Sm (100)	La (25)			36.4 Ce (12)	Au (5)
	97.6 Ti (40)	Sm (15)			35.0 Yt (150)	Sm (25)
(800)Ba	96.9 Ru (25)	Fe (10)			34.3 Ce (12)	Zr (6)
	95.5 Nd (8)	U (2)			33.2 Cb (30)	Sm (15)
(400)Fe	94.9 Gd (20)	Pr (4)			32.7 Yb (30)	Nd (12)
(80)Ca	93.7 Eu (40)	Co (25)			31.9 Sm (50)	Sm (50)
	92.3 Er (12)	Nd (12)		(100)Fe	30.7 Ta (150)	Cb (80)
	91.7 Mn(100)	Pr (20)			29.9 Co (50)	Pr (8)
	90.3 Co (70)	Sm (60)			28.2 Eu (200)	Sm (50)
	89.1 Yb (800)	Zr (50)			27.5 Cu (3)	Dy (2)
	88.6 Sm (4)	V (2)			26.6 Sm (100)	Ta (5)
	87.6 Sm (60)	Nd (5)			25.7 Nd (15)	Sm (10)
	86.5 Pr (30)	Ir (8)			24.3 Mo(100)	Gd (10)
	85.3 Ta (500)	La (50)			23.1 Er (6)	
	84.5 Sm (100)	Zr (6)			22.4 Gd (50)	Cb (6)
	83.0 Eu (30)	Gd (10)		(60)Fe	21.3 Cr (35)	Co (20)
(100)Ba	82.9 Ni (35)	Nd (10)			20.3 Sc (2)	Band
	81.8 Fe (12)	Ce (6)			19.9 Fe (18)	Ti (15)
	80.1 Gd (40)	Nd (10)			18.9 Mo(5)	Ir (5)
	79.1 Zn (10)	Nd (6)		(100)Sm	17.8 Co (200)	Yb (125)
	78.0 Pr (15)	Sm (6)			16.5 F (100)	Band
	77.8 Co (80)	Lu (30)			15.3 F (100)	Band
	76.2 Bi (8)	Eu (6)			14.1 F (100)	Band
	75.6 Fe (8)	Bi (8)			13.9 Mn(25)	Sc (10)
	74.5 Co (25)	Sm (25)			12.3 Mo(15)	Rh (8)
	73.9 Mo(20)	Sm (8)		(100)Fe	11.3 Eu (500)	Re (20)
	72.3 Sm (150)	Cs (15)			10.1 Eu (500)	La (100)
	71.2 Mo(50)	Ca (40)			09.1 Mo(25)	Hf (3)
	70.4 Sm (60)	Zr (40)		(50)Fe	08.0 Sr (50)	Sm (10)
	69.2 Fe (8)	Sm (7)			07.6 Nd (5)	
	68.4 La (25)	Sm (4)			06.1 Eu (100)	Sm (30)
	67.4 Ce (20)	Pr (6)			05.9 Tb (6)	Er (4)
	66.5 Gd (25)	Ce (5)			04.2 W (25)	Sm (5)
	65.0 U (25)	Nd (8)			03.1 Os (15)	Tb (4)
	64.9 Cd (5)	Cb (4)			02.0 Yt (12)	W (5)
	63.1 Lu (400)	Co (25)			01.4 Tm(40)	Mo(20)
(125)Ca	62.5 Co (60)	Gd (50)		(200)Fe	00.9 Eu (700)	Yb (200)
	61.8 Ce (5)	Sm (5)			**6300**	
	60.2 Tm(400)	Er (6)			99.0 La (15)	W (5)
	59.9 Ta (15)	Sm (15)			98.2 Sm (12)	Pt (6)
	58.0 Ce (25)	Co (2)			97.9 Pr (12)	U (12)
	57.9 Eu (500)	Th (15)			96.5 Co (10)	Ce (6)
	56.2 Sm (30)	Gd (15)			95.1 Co (125)	U (100)
	55.9 La (100)	Sm (40)			94.2 La (150)	Nd (4)
	54.9 Co (200)	La (125)		(100)Fe	93.6 Pr (25)	Ce (15)
	53.4 Pr (10)	Mo(9)			92.7 U (20)	Ta (15)
	52.3 V (25)	Sm (25)			91.1 Mo(12)	Mn(3)
	51.1 Co (70)	Zr (10)			90.8 Sm (100)	La (70)
(100)Ba	50.2 Co (1000)Ta (200)				89.8 Sm (100)	Ta (100)
(60)Ca	49.1 U (100)	Cb (6)			88.2 Sr (35)	Er (12)
	48.1 La (10)	Sm (8)			87.0 Yt (8)	Ta (3)

	86.5 Sr (35)	Cs (25)		24.4 Eu (10)	V (8)
	85.1 Nd (100)			23.5 Mo (12)	Dy (2)
	84.6 Mn (25)	Nd (6)		22.3 Pr (30)	Fe (8)
	83.8 Eu (350)	U (8)		21.8 Re (100)	Sm (60)
	82.7 Eu (200)	Gd (60)		20.4 Co (80)	La (70)
	81.4 Ti (10)	V (2)		19.5 Rh (50)	Nd (5)
	80.9 Gd (100)	Sr (30)	(50) Ti	18.0 Fe (40)	Pt (30)
	79.6 U (15)	V (8)		17.1 Gd (10)	Sr (4)
	78.9 Mn (20)	Ni (20)		16.8 Mo (4)	Ru (4)
	77.6 Pr (5)	Au (5)		15.7 Sm (15)	Mn (5)
	76.9 Th (12)	Ru (9)		14.6 Ni (300)	Co (50)
	75.9 U (5)	Nd (5)		13.0 Zr (200)	Co (50)
	74.1 La (6)	Er (4)	(80) Ti	12.2 Au (8)	Cb (6)
	73.0 Ta (50)	Eu (5)		11.2 Co (30)	V (10)
	72.4 U (50)	Ho (10)		10.4 Nd (50)	Ce (20)
	71.9 Sm (15)	Ce (10)		09.5 Ta (100)	Ta (15)
	70.8 Mn (4)	Ca (4)		08.1 Yb (20)	Er (15)
	69.2 Eu (300)	Sr (25)		07.7 Re (100)	Sm (60)
	68.2 Sm (50)	Gd (10)		06.6 Ce (10)	F (100) Band
	67.4 Sm (10)	U (2)		05.1 Gd (100)	Sc (80)
(80) Ti	66.3 Ni (15)	Lu (15)		04.8 F (150)	Band
	65.5 Nd (15)	Lu (10)	(200) Ti	03.3 Eu (700)	F (150) Band
	64.9 Ti (8)	W (4)		02.3 Sm (25)	F (150) Band
	63.9 Sr (25)	Ru (4)	(50) Fe	01.5 Sm (50)	Mo (15)
	62.3 Zn (1000)	Cr (150)		00.9 F (150)	Band
	61.8 Sc (10)	Nd (10)		**6200**	
	60.8 Ta (100)	La (15)		99.7 Eu (500)	Rb (300)
	59.0 Pr (40)	U (30)		98.3 Rb (1000)	Nd (20)
	58.1 La (10)	Cu (8)		97.0 Nd (15)	Fe (10)
	57.1 Sm (50)	Mo (40)		96.0 La (50)	W (30)
	56.1 Ta (100)	Mn (8)		95.5 Ce (20)	Sm (10)
	55.8 Eu (200)	Fe (15)		94.1 Rh (3)	Nd (2)
	54.8 Lu (20)	Cs (15)		93.5 La (30)	U (15)
	53.4 Sm (50)	Sm (20)		92.0 V (50)	W (30)
	52.6 Tm (40)	Co (2)		91.3 Eu (250)	Sm (100)
	51.4 Co (25)	Er (6)		90.7 Mo (15)	Fe (5)
	50.0 Eu (600)	Re (100)		89.9 Sm (60)	Ta (20)
	49.7 Mn (15)	V (15)		88.3 Eu (300)	Ir (20)
	48.9 Cb (8)	Sm (4)		87.9 Ta (10)	La (5)
	47.8 Co (125)	Mg (10)		86.4 Re (20)	Cb (10)
	46.0 Ta (40)	Gd (30)		85.1 V (50)	W (30)
	45.3 Lu (60)	Sr (25)		84.1 Sm (10)	Sc (4)
	44.1 Mn (20)	Fe (5)		83.4 Pt (10)	Ta (4)
	43.9 Ce (15)	Pr (10)		82.6 Co (300)	V (15)
	42.8 Th (15)	Sc (2)		81.3 Ta (50)	Pr (25)
(90) Ba	41.6 Ta (50)	Nd (15)		80.1 U (10)	Sc (5)
	40.0 Sm (25)	Co (10)		79.1 Th (20)	Hf (15)
	39.1 Ni (50)	V (25)		78.1 Au (700)	Re (20)
	38.8 Sm (5)	Yt (5)		77.4 Rh (15)	Th (10)
	37.2 Ce (8)	Co (5)		76.6 Co (40)	Ce (10)
(80) Ti	36.1 Fe (60)	Gd (10)		75.1 Co (25)	Cb (6)
(50) Fe	35.7 Eu (200)	Ce (15)		74.7 Yb (100)	V (50)
	34.4 Ir (20)	Tb (6)		73.0 Co (70)	Sm (15)
	33.8 La (3)	Nd (2)		72.0 Ce (20)	Sr (15)
	32.9 Ta (50)	Sc (20)		71.4 Re (15)	Sm (8)
	31.3 Gd (15)	Tb (8)		70.2 Ce (8)	Nd (5)
	30.1 Cr (200)	Ru (20)		69.4 Os (8)	Ce (4)
	29.9 Cd (30)	Sm (16)		68.7 Ta (200)	Cu (40)
	28.8 Sm (10)	Nd (3)		67.2 Sm (150)	Zr (10)
	27.4 Sm (100)	Ni (25)		66.9 Eu (120)	La (80)
	26.5 Pt (50)	V (15)		65.8 Mo (25)	Fe (12)
(40) Ti	25.1 Cd (100)	Ta (100)		64.2 Mo (15)	Eu (9)

Wave-length Table-chart

```
             63.4 Eu (10)   Pt  (4)              01.7 Nd (15)   Sm (5)
             62.2 Eu (600)  La (125)             00.3 Fe (15)   Th (5)
    (300)Ti  61.0 V  (20)   Th (15)                   6100
             60.7 Cb (20)   Re (20)              99.1 V  (100)  Lu (40)
             59.0 Dy (15)   Sm (8)               98.3 Sm (20)   Cd (15)
    (300)Ti  58.7 Ti (200)  V  (35)              97.6 Mo (15)   Co (5)
           * 57.5 Co (70)   Zr (30)              96.3 Sb (6)    W  (5)
             56.3 Ni (600)  Ta (300)             95.0 Eu (300)  Re (15)
             55.8 Sm (10)   Ho (10)              94.4 Sm (10)   Tb (4)
             54.2 Si (25)   Si (15)              93.1 Ta (30)   Co (15)
             53.7 Rh (35)   Ce (8)               92.6 Sm (50)   Zr (20)
    (60)Fe   52.5 Re (2)    Sm (2)      (100)Fe  91.1 Ni (500)  Sm (15)
             51.8 V  (70)   Cb (30)              90.6 Cr (5)    Yb (3)
             50.4 Eu (70)   Nd (10)              89.6 Ta (50)   V  (10)
             49.9 La (300)  Co (125) Ta (100)    88.0 Eu (500)  Co (200)
             48.9 Hf (80)   Nd (15)              87.9 Sm (30)   W  (5)
             47.2 Co (8)    Yb (3)               86.1 Ti (35)   Rh (30)
             46.7 Sm (80)   Yb (40)              85.1 Hf (10)   Fe (2)
             45.6 Sc (6)    Eu (5)               84.7 Th (4)    Bi (3)
             44.2 Sm (40)   Nd (35)              83.9 Nd (20)   Er (6)
             43.2 Re (50)   V  (35)              82.8 Sm (60)   Th (12)
             42.3 Lu (40)   V  (20)              81.0 Sm (50)   Tm (40)
             41.1 Pr (8)    Ce (6)               80.4 Gd (30)   Fe (6)
             40.1 Li (300)  V  (20)              79.8 Sm (30)   Ta (30)
             39.4 Sc (8)    Zn (8)               78.7 Eu (120)  Nd (15)
             38.5 La (25)   Nd (20)              77.4 Sm (5)    Ni (5)
             37.6 Sm (50)   Ce (10)              76.8 Ni (400)  Sm (8)
             36.7 La (15)   Sm (15)              75.4 Ni (300)  Ce (8)
             35.3 Lu (25)   Pb (20)              74.9 Sm (30)   Sm (10)
             34.8 La (25)   Ho (10)              73.0 Eu (600)  Fe (18)
             33.7 Eu (100)  V  (30)              72.7 La (15)   Pt (15)
             32.4 Co (25)   Th (10)              71.8 U  (30)   Sn (10)
             31.7 Sm (4)                         70.4 Nd (30)   V  (30)
    (60)Fe   30.9 Co (200)  V  (70)      (40)Ca  69.5 Ca (25)   Pb (10)
             29.4 Re (40)   Dy (3)               68.3 Sm (30)   Dy (6)
             28.9 Ce (10)   Lu (10)              67.3 Ta (8)    La (3)
             27.7 Os (30)   Fe (5)       (15)Ca  66.6 Nd (20)   Sm (8)
             26.5 Nd (20)   Sm (5)               65.6 La (100)  Pr (50)
             25.2 Ru (20)   Sm (18)              64.5 Sm (20)   Cb (10)
             24.5 V  (50)   Zr (6)               63.4 Ni (100)  Ca (10)
             23.9 Ni (30)   Nd (25)      (40)Ca  62.1 Co (60)   Nd (8)
             22.1 U  (8)    Nd (8)               61.1 Pr (50)   Ca (10)
    (80)Ti   21.8 Lu (500)  Cb (20)              60.7 Na (500)  Sm (40)
    (100)Ti  20.4 Sm (10)   Cu (8)               59.6 Rb (400)  Lu (50)
    (40)Fe   19.2 Cb (6)    Ti (5)               58.8 Ta (80)   Mo (10)
             18.3 V  (15)   La (8)               57.5 Sm (30)   Nd (25)
             17.9 Re (40)   Mo (20)              56.9 Sm (30)   Si (10)
             16.3 V  (60)   Ce (10)              55.3 Si (50)   Si (20)
    (100)Ti  15.2 Pt (20)   U  (12)              54.2 Na (500)  Ta (200)
             14.6 Zr (18)   Mo (7)               53.7 W  (15)   Eu (8)
             13.8 V  (50)   Fe (20)              52.5 Yb (60)   Ta (60)
             12.8 Cs (100)  Ti (18)              51.1 Gd (15)   Th (10)
             11.1 Co (25)   Ir (15)              50.1 V  (40)   Yb (30)
             10.6 Sc (20)   Nd (10)              49.1 Sm (60)   Ti (30)
             09.4 Mo (8)    Ce (5)               48.1 Cb (20)   Yt (100) Band
             08.9 Ce (12)   Mo (12)              47.3 Cu (20)   Ta (15)
             07.2 V  (20)   Th (10)      (400)Ti 46.2 Re (50)   La (20)
             06.3 Rb (800)  Sm (25)              45.8 Re (50)   Si (15)
             05.8 Sm (5)    Co (3)               44.5 Ta (50)   Re (25)
             04.6 Ni (10)   Sb (6)               43.2 Zr (300)  Sm (30)
             03.2 Re (25)   Sm (20)              42.9 La (50)   Cb (10)
             02.8 Hf (2)    Ce (2)       (2000)Ba 41.7 Fe (10)  Pr (6)
```

	40.4 Zr (40)	Ta (40)		79.5 Sb (20)	Sc (100)Band
	39.3 Sm (6)	Eu (6)		78.4 Fe (8)	
	38.3 Ti (15)	Mo(10)		77.3 V (300)	Eu (100)
(100)Fe	37.6 Fe (10)	Ce (4)		76.8 Pt (10)	Er (10)
(100)Fe	36.6 Gd (5)	La (2)		75.5 Eu (300)	Sm (8)
	35.3 V (30)	Sm (20)		74.2 Eu (10)	Sm (5)
	34.5 Zr (300)	La (70)		73.9 Sb (20)	Nd (20)
	33.6 Ho (10)	Nd (5)		72.7 Sc (100)	Band
	32.1 Yt (200)	Band		71.7 Nd (20)	Sm (2)
	31.1 Sm (10)	Si (5)		70.7 Rb (600)	Sm (40)
	30.1 Ni (15)	Mo(12)		69.4 Ce (20)	Nd (4)
	29.5 La (40)	Co (20)		68.9 Sn (12)	La (10)
	28.7 Sm (20)	W (15)		67.7 Sm (40)	Ir (25)
	27.4 Zr (500)	Cu (80)		66.0 Nd (20)	Ce (8)
(150)Ti	26.2 La (20)	Sm (10)	(50)Fe	65.4 W (7)	Nd (4)
	25.8 Sm (15)	Re (10)	(80)Ti	64.4 F (200)	Band
	24.6 Eu (150)	Zr (40)	(200)Ba	63.1 V (10)	Mo(6)
	23.6 Sm (50)	Ce (15)		62.8 Zr (30)	F (150)Band
(100)Ca	22.6 Co (125)	Co (40)		61.6 Eu (15)	Sm (10)
	21.9 Zr (60)	Ti (35)		60.4 F (100)	Band
	20.5 Th (15)	Zn (12)		59.7 Pb (20)	Yb (10)
	19.5 V (30)	Cu (25)		58.1 V (60)	Mo(12)
	18.7 Eu (400)	Pr (6)		57.3 Eu (600)	Ce (15)
	17.7 Sm (8)	Tb (8)		56.0 Fe (10)	Cb (10)
	16.1 Ni (150)	Co (80)		55.0 Lu (150)	Pr (100)
	15.5 W (12)	Er (4)		54.8 Mo(20)	Sn (12)
	14.1 Re (30)	Gd (25)		53.6 Ta (150)	Sm (5)
	13.4 Nd (8)	Pt (5)		52.9 Eu (8)	Yb (8)
	12.8 Th (15)	Sm (10)		51.7 U (8)	Eu (6)
	11.5 Cd (100)	La (50)		50.8 Mo(10)	U (8)
(200)Ba	10.7 Sm (40)	Ir (35)		49.5 Eu (1000)	Co (50)
	09.9 Sc (40)	Band		48.4 Yb (15)	Cb (6)
	08.1 Ni (200)	Eu (150)		47.2 Ta (150)	Mo(15)
	07.4 Eu (30)	La (25)		46.6 Pr (12)	Ru (4)
	06.1 Gd (15)	V (15)		45.3 Ta (200)	Sm (40)
	05.4 Co (10)	Cb (10)		44.6 Eu (200)	Sm (60)
	04.8 Sm (30)	Th (12)		43.3 Ce (30)	W (10)
	03.6 Li (2000)	Sm (30)		42.8 Sm (15)	Sm (10)
(80)Ca	02.7 Rh (100)	Fe (15)		41.3 Sm (40)	Lu (20)
	01.5 Ta (150)	Mo(40)		40.8 Eu (6)	Ce (4)
	00.3 La (30)	Eu (12)		39.7 V (100)	Mo(12)
	6000			38.6 La (50)	Tb (8)
	99.3 Eu (600)	Cd (300)		37.7 Sn (12)	Th (8)
(60)Ti	98.6 Ce (10)	Hf (5)		36.2 Sc (200)	Band
	97.6 Sb (15)	Er (4)		35.5 Sm (4)	Ce (4)
	96.7 Sm (15)	Mo(4)		34.0 Cs (35)	Nd (20)
	95.2 Sm (4)	Fe (3)		33.2 Sm (25)	Sm (20)
	94.0 Sm (5)	Tb (4)		32.6 Zr (25)	Cu (8)
	93.1 Co (200)	Ce (10)		31.3 Cd (30)	Nd (25)
	92.8 Ti (35)	Ta (30)		30.6 Mo(300)	Yt (100)Band
(125)Ti	91.1 Sm(20)	Tb (4)		29.0 Eu (500)	Cb (10)
	90.2 V (60)	Ta (50)		28.3 W (7)	Zr (6)
	89.6 Sm (5)	Fe (4)		27.2 Mo(20)	Sm (15)
	88.6 Sm (10)	Ce (8)		26.0 Ir (20)	Pt (20)
	87.5 Pr (15)	Th (12)		25.4 Mo(25)	Pr (25)
	86.2 Ni (100)	Co (80)		24.1 Ce (50)	Fe (20)
(100)Ti	85.2 Pr (8)	Th (6)		23.1 Eu (200)	Sm (15)
	84.1 Sm (50)	Ta (15)		22.7 Au (15)	Er (8)
	83.8 Eu (500)	Ba (10)	(300)Fe	21.8 Mn(80)	Co (50)
	82.4 Co (300)	Sm (15)		20.7 Ta (300)	Fe (10)
	81.4 V (100)	W (25)	(150)Ba	19.4 Pr (8)	Yt (150)Band
	80.6 Gd (25)	Ce (6)		18.1 Eu (1000)	Ti (15)

Wave-length Table-chart

	17.3 Sm (50)	Pr (40)		56.9 Au (35)	Fe (12)
(100)Fe	16.6 Mn(80)	Co (40)		55.9 Ho (100)	Sm (25)
	15.5 Eu (150)	Ta (40)		54.2 Eu (30)	Re (3)
	14.8 Er (8)	Mn(8)	(150)Ti	53.1 Eu (40)	Eu (30)
(100)Fe	13.5 Mn(100)	Co (30)		52.7 Fe (8)	Sc (5)
	12.5 Eu (700)	W (30)		51.7 Ta (15)	Tb (15)
	11.2 Sm (60)	Pb (25)		50.6 Ce (8)	Nd (8)
	10.3 Cs (50)	Gd (10)		49.0 Tl (20)	Nd (10)
	09.8 Ta (40)	W (6)		48.0 Ho (200)	Si (50)
	08.5 Fe (18)	Gd (15)		47.5 W (15)	Nd (8)
	07.3 La (70)	Co (50)		46.4 Co (70)	Sm (15)
	06.3 Co (50)	Ce (15)		45.8 Dy (5)	Yt (4)
	05.6 Eu (60)	Ce (20)		44.0 Ta (80)	Ti (10)
	04.5 Lu (400)	Eu (300)		43.2 Re (100)	Nd (15)
(30)Fe	03.6 Yt (200)	Band		42.7 Eu (30)	Ce (10)
	02.6 V (30)	Ho (8)	(100)Ti	41.7 Pr (5)	Sm (5)
	01.8 Pb (40)	Sm (50)		40.8 Ce (40)	Tb (15)
	00.6 Co (80)	Ce (6)		39.7 Ta (80)	Yt (100) Band
	5900			38.8 Sm (50)	Th (8)
(70)Ti	99.6 Ti (15)	Fe (10)	(60)Ti	37.8 Mo(20)	Ce (20)
	98.9 Ti (8)	Sm (3)		36.1 Sm (35)	La (15)
(150)Ba	97.2 Ta (200)	Cb (50)		35.3 Co(150)	La (20)
	96.0 Ti (20)	Nd (15)		34.6 Fe (15)	Tb (10)
	95.9 Os (50)	Re (20)		33.7 Ho (200)	Er (8)
	94.6 Sm (60)	Nd (15)	(80)Ti	32.1 Sm (50)	Sm (45)
	93.8 Sm (20)	Ru (15)		31.6 Ta (40)	Ta (40)
	92.8 Eu (1000)	C (100) Band		30.6 La (250)	Fe (30)
	91.8 Co (900)	Yb (50)		29.8 Ce (13)	Th (5)
	90.0 Mo(10)	Nd (8)		28.8 Mo(100)	Ce (25)
	89.0 Th (20)	Nd (15)		27.8 Fe (10)	Cb (5)
	88.1 Mo(30)	Ti (18)		26.5 Eu (80)	Mo(70)
(25)Fe	87.6 Yt (300)	Band		25.1 Zr (20)	Ta (20)
	86.1 Pr (25)	U (25)		24.5 V (250)	Sm (10)
	85.5 Ba (3)	Sm (2)		23.3 Sm (15)	Mo(8)
(50)Fe	84.8 Lu (40)	Ti (15)	(100)Ti	22.1 Tb (10)	Nd (5)
(35)Fe	83.5 Rh (200)	Lu (30)		21.7 Ho (200)	Ru (25)
	82.9 Ho (200)	Mo(20)		20.4 Ce (15)	Sm (15)
	81.2 Pr (12)	Ho (8)		19.3 Sm (60)	Ru (20)
	80.7 V (50)	Eu (10)	(80)Ti	18.5 Ta (80)	Rh (20)
	79.3 Sm (15)	Ce (5)		17.6 La (25)	Sm (3)
(125)Ti	78.5 V (100)	W (7)		16.5 Ta (30)	Fe (25)
	77.2 Gd (8)	Nd (3)		15.5 Co (200)	U (125) Eu (200)
	76.3 U (50)	W (5)	(50)Fe	14.1 Th (12)	Er (8)
	75.9 Ce (30)	V (18)		13.8 Sm (20)	Gd (15)
	74.2 Mo(20)	Hf (10)		12.9 Ce (20)	Sm (15)
	73.0 Yt (600b)	Ho (125)		11.4 Gd (3)	Re (2)
(100)Ba	72.7 Eu (300)	W (25)		10.1 Ce (15)	Tb (10)
(150)Ba	71.7 U (50)	Eu (40)		09.9 Eu (30)	Er (20)
	70.8 Sm (15)	Sn (10)		08.3 Yb (20)	Fe (8)
	69.7 Re (20)	Fe (5)		07.6 Ba (20)	Rh (5)
	68.8 Sm (40)	Er (12)		06.0 Er (20)	Ce (10)
	67.1 Eu (2000)	Tb (40)		05.6 Fe (12)	Nd (5)
	66.0 Eu (1000)	Er (8)		04.7 Tb (15)	Ho (12)
(150)Ti	65.8 Sm (125)	W (25)		03.3 Ti (40)	Sm (20)
	64.7 Ba (7)	Ce (5)		02.0 Er (30)	Eu (30)
	63.7 Eu (60)	Sm (40)		01.9 Ta (80)	Mo(30)
	62.7 Au (35)	Ho (12)		00.6 Cb (200)	Ho (8)
	61.0 Tb (25)	Th (8)		**5800**	
	60.1 Sm (20)	La (10)	(150)Ti	99.3 Tb (15)	Mo(12)
	59.6 Ce (6)	Pr (3)		98.9 Sm (25)	Tb (25)
	58.9 Er (20)	Yb (10)		97.3 Sm (100)	Yb (7)
	57.5 Sm (40)	Cb (8)		96.7 La (80)	Band

	95.9 Na (5000)	Tm (80)	
	94.8 La (25)	Ir (20)	
	93.3 Mo (70)	Cb (15)	
	92.5 Ho (50)	Mo (20)	
	91.2 Eu (200)	Mo (25)	Nd (20)
	90.4 Hf (8)	Co (7)	
	89.9 Na (9000)	Mo (50)	
	88.3 Mo (150)	Cr (20)	
	87.9 Nd (25)	Ho (12)	
	86.4 Er (40)	Gd (6)	
	85.6 Zr (25)	Th (10)	
	84.4 Cr (18)	Mo (12)	
	83.8 Fe (15)	Nd (15)	
	82.9 Ho (200)	Ta (80)	
	81.1 Er (30)	Mo (20)	
(60) Ti	80.3 La (30)	W (15)	
	79.7 Zr (60)	Pr (10)	
	78.0 Sm (20)	Sm (15)	
	77.3 Ta (100)	Ti (15)	
	76.5 Mo (25)	Cb (10)	
	75.3 Fe (15)	W (8)	
	74.1 Sm (40)	Cb (30)	
	73.1 Tb (10)	Fe (8)	
	72.9 Eu (300)	Er (20)	
	71.0 Sm (25)	Ce (20)	
	70.6 Tb (25)	Ho (20)	
	69.3 Mo (50)	Fe (10)	
	68.6 Sm (40)	Mo (20)	
	67.7 Sm (50)	Ca (8)	
(300) Ti	66.4 Ta (60)	Cb (50)	
	65.8 Ta (40)	Nd (15)	
	64.7 Eu (20)	W (20)	
	63.7 La (40)	Sm (12)	
	62.3 Fe (35)	Au (30)	
	61.2 Tb (20)	Mo (20)	
	60.2 Ho (200)	Sm (40)	
	59.6 Fe (15)	Pr (15)	
	58.2 Mo (200)	C (400)	Band
(40) Ca	57.7 Os (80)	Ni (50)	
	56.6 W (15)	Pr (10)	
	55.2 Er (30)	La (30)	
	54.5 Yb (30)	Tb (10)	
(300) Ba	53.6 U (10)	Ce (9)	
	52.0 Re (40)	Tb (15)	
	51.5 Mo (40)	Tb (40)	
	50.3 V (40)	Er (20)	
	49.6 Ta (140)	Mo (70)	
	48.8 Mo (50)	·Sm (30)	
	47.1 Pr (10)	Ni (5)	
	46.2 V (10)	Tb (10)	
	45.7 Eu (30)	U (20)	
	44.8 Pt (40)	Cs (30)	
	43.9 Ta (15)	Sm (8)	
	42.2 Hf (50)	Sm (40)	
	41.9 Sm (10)	Er (8)	
	40.1 Pt (80)	Gd (10)	
	39.4 Ho (30)	Mo (15)	
	38.6 Cb (200)	W (25)	
	37.3 Au (400)	Yb (50)	
	36.3 Sm (100)	U (30)	
	35.8 Ce (25)	Mo (20)	
	34.3 Re (200)	Yb (60)	
	33.9 Er (12)	W (12)	
	32.0 K (50)	Tb (10)	
	31.0 Eu (2000)	Rh (80)	
	30.7 V (100)	Sm (20)	
	29.7 Sm (35)	La (25)	
	28.0 Ru (6)	Fe (4)	
	27.5 La (15)	Tb (15)	
(150) Ba	26.2 Er (50)	Nd (15)	
	25.8 Nd (25)	Mo (35)	
	24.0 Nd (5)	Fe (3)	
	23.7 Pr (60)	Ti (35)	
	22.9 Ce (12)	Pr (8)	
	21.9 La (40)	Ho (12)	
	20.6 Sm (40)	Eu (30)	
	19.4 Cb (20)	Yb (7)	
	18.7 Eu (1000)	Pr (10)	
	17.5 V (100)	V (50)	
	16.5 Ta (40)	Fe (15)	
	15.8 Re (50)	Mo (20)	
	14.8 Sm (60)	Ru (25)	
	13.8 Nd (30)	Ti (8)	
	12.9 Ce (40)	K (30)	
	11.1 Ta (100)	Nd (15)	
	10.7 Ce (15)	Pr (10)	
	09.5 Hf (20)	Tb (15)	
	08.3 La (25)	Re (6)	
	07.1 V (75)	Gd (6)	
	06.9 Rh (100)	Mo (20)	
(70) Ba	05.6 La (60)	Ni (50)	
(100) Ti	04.2 Nd (100)	W (25)	
	03.1 Tb (40)	Yb (15)	
	02.8 Sm (80)	Mo (30)	
	01.9 K (50)	Sm (25)	
(100) Ba	00.2 Eu (200)	Sm (80)	

5700

	99.8 V (40)	W (12)	
	98.5 U (35)	Ho (12)	
	97.5 La (80)	Zr (50)	
	96.5 W (20)	Th (10)	
	95.9 F (100)	Band	
	94.2 Cb (15)	Ce (4)	
	93.5 C (30)	F (100)	Band
	92.7 Rh (40)	Eu (30)	
	91.3 La (200)	Mo (100)	
	90.3 F (100)	Band	
	89.2 La (125)	Fe (8)	
	88.2 Nd (30)	Sm (30)	
	87.5 Cb (80)	F (100)	Band
	86.9 Sm (200)	V (75)	
(100) Ti	85.9 Tb (40)	Cr (20)	
	84.3 V (50)	F (100)	Band
	83.7 Eu (150)	Cr (30)	
	82.1 Cu (1000)	F (150)	Band
	81.8 Sm (100)	Cr (20)	
	80.7 Ta (80)	Ta (60)	
	79.4 F (200)	Band Pr (50)	
	78.3 Sm (50)	Mo (12)	
(500) Ba	77.6 Zn (10)	F (100)	Band
	76.8 Re (300)	Ta (80)	
	75.4 Lu (50)	Fe (12)	
(70) Ti	74.0 Mo (20)	F (100)	Band
	73.7 Sm (100)	Ce (30)	

Wave-length Table-chart

	72.4 V (50)	Si (30)		10.8 Er (20)	Sm (10)
	71.6 Yb (30)	F (100) Band	(100) Fe	09.3 Ni (100)	In (50)
	70.0 La (25)	Nd (20)	(30) Ti	08.2 Nd (60)	Si (40)
	69.5 Hg (600)	La (30)		07.6 Pr (100)	Th (20)
	68.8 Ce (15)	Ho (8)		06.9 V (200)	Ho (50)
	67.9 Ta (100)	Hf (15)		05.7 Mo(40)	Fe (15)
(70) Ti	66.5 Ta (80)	Ho (8)		04.3 Ta (40)	V (10)
	65.2 Eu (2000)	Os (15)		03.5 V (200)	La (20)
	64.2 Tm(50)	Cb (10)	(60) Ti	02.6 Nd (25)	Cr (20)
(80) Fe	63.0 Pt (30)	Sm (20)	(50) Fe	01.5 Gd (20)	Si (15)
(70) Ti	62.2 Er (30)	Tb (15)		00.2 Sc (400)	Cu (350)
	61.8 La (60)	V (25)		**5600**	
	60.8 Ni (50)	Cb (30)		99.0 Ru (125)	Ta (80)
	59.5 Sm (60)	W (12)		98.5 V (300)	Cr (30)
	58.0 Tm(15)	U (10)		97.8 W (35)	Gd (6)
	57.6 Er (30)	Sm (20)		96.1 La (50)	Ce (40)
	56.1 Pr (25)	Sm (15)		95.0 Pd (50)	Ce (10)
	55.8 Ta (40)	Yb (15)		94.9 Ni (40)	Cr (35)
	54.6 Ni (150)	Si (40)		93.6 Ho (12)	Ru (7)
	53.1 Fe (40)	Cr (15)		92.2 Pb (20)	Ce (25)
	52.9 Re (200)	Er (12)		91.4 Ho (200)	Pr (20)
	51.4 Mo(125)	Ho (30)		90.4 Si (25)	Pr (10)
	50.6 V (50)	W (7)	(80) Ti	89.4 Mo(80)	Mo(12)
	49.1 Nd (30)	Th (12)	(300) Na	88.2 Nd (150)	Ta (100)
	48.1 Zr (100)	Band Ni(40)		87.7 V (15)	Tb (15)
	47.5 Tb (60)	Mo(15)		86.8 Sc (200)	Rh (100)
	46.7 Ta (60)	Cr (12)		85.7 Tb (40)	Ho (8)
	45.9 Ru (10)	Sm (10)		84.2 Eu (125)	Tm(40)
	44.4 La (80)	Nd (25)		83.2 V (50)	Cr (3)
	43.4 V (60)	Ce (25)		82.6 Na (80)	Ni (50)
	42.0 Nd (40)	Bi (30)		81.0 Eu (40)	Ho (20)
	41.1 Sm (60)	Mo(10)	(60) Ba	80.1 Zr (50)	Os (20)
	40.6 La (100)	Sm (80)	(50) Ti	79.9 Sm (15)	Ru (10)
(70) Ti	39.5 Er (30)	Nd (15)		78.4 Tb (10)	Dy (3)
	38.9 Eu (300)	Sm (15)		77.8 Mo(25)	Ce (25)
	37.0 V (100)	Cb (5)		76.0 Tb (15)	Ho (12)
	36.5 Lu (150)	Sm (40)	(90) Ti	75.7 Na (150)	Tm(100)
	35.0 W (50)	Zr (25)		74.7 Ho (200)	Mo(30)
	34.0 V (35)	Mo(20)		73.8 Eu (200)	Mo(20)
	33.8 Gd (20)	Er (12)		72.0 Mo(10)	Sm (5)
	32.9 Sm (100)	Th (10)		71.8 Sc (300)	Cb (200)
	31.2 V (250)	Fe (10)		70.8 V (150)	Na (100)Pd (100)
	30.8 Eu (300)	C (150) Band		69.9 Ce (50)	Nd (40)
	29.9 Sm (40)	Nd (30)		68.3 V (75)	Pr (25)
	28.2 In (50)	Mo(15)		67.9 Re (100)	Tb (25)
	27.0 V (150)	V (75)		66.0 Sm (9)	Er (8)
	26.8 Au (35)	Nd (25)		65.6 Cb (100)	Si (20)
	25.6 V (40)	Sm (40)		64.7 Cb (100)	Ta (60)
	24.4 Rb (600)	Rb (50)		63.8 Cs (15)	Ce (12)
	23.0 W (15)	U (15)	(140) Ti	62.1 Fe (50)	Yt (20)
	22.7 Mo(80)	Tb (10)		61.4 Sm (25)	Pr (10)
	21.9 Os (80)	Sm (25)		60.5 Er (8)	Rh (4)
	20.0 Yb (300)	Sm (30)		59.8 Sm (60)	Co (25)
	19.1 Sm (60)	Hf (40)	(100) Fe	58.8 Fe (30)	Tm(20)
	18.1 Zr (150)	Band		57.4 V (150)	La (50)
	17.9 Sm (30)	Sc (20)		56.3 Sm (15)	V (13)
	16.2 V (60)	Ti (40)		55.1 Ce (40)	Au (35)
(70) Ti	15.1 Ni (50)	Ta (30)		54.6 Eu (8)	Th (8)
	14.0 La (15)	Re (6)		53.7 Rb (200)	Re (10)
	13.9 Ti (25)	Ba (10)		52.0 Tb (15)	Sm (10)
	12.4 La (20)	Cr (15)		51.9 Yb (50)	Al (10)
(50) Ti	11.7 Sc (100)	Ni (50)		50.1 Mo(90)	Ce (15)

	49.6 Ni (15)	Cr (9)			
(80)Ti	48.0 Rb (400)	La (150)	(35)Ca	88.3 La (40)	Yb (30)
	47.2 Co (600)	Sm (10)		87.8 Ni (50)	Th (20)
	46.1 V (150)	Ce (10)	(400)Fe	86.7 Eu (300)	Eu (200)
	45.8 Eu (1000)	Ta (80)		85.6 Ti (25)	Zn (12)
(150)Ti	44.1 Sm (40)	Yt (15)		84.0 Ta (60)	Os (50)
	43.3 Sm (20)	Gd (20)		83.6 Gd (25)	Tl (15)
	42.1 Cb (80)	W (15)		82.7 Ce (20)	Ce (3)
	41.8 Ni (25)	Fe (15)	(20)Ca	81.8 Yt (100)	Ti (25)
	40.6 Ho (100)	Ta (25)		80.0 Eu (300)	Os (8)
	39.5 Nd (25)	La (15)		79.6 Eu (200)	Sm (10)
	38.2 Fe (40)	Pr (30)		78.7 Rb (150)	Ni (50)
	37.2 Sm (80)	Co (20)		77.1 Eu (1000)	Yt (15)
	36.2 Ru (100)	Co (20)	(150)Fe	76.1 Cb (80)	Gd (12)
	35.5 V (35)	Nd (25)		75.1 Mo(20)	Th (10)
	34.8 Mo(30)	U (10)		74.9 Sm (50)	Sm (15)
	33.9 Fe (20)	Ce (15)		73.4 Sm (80)	Re (30)
	32.5 Eu (200)	Mo(100)	(300)Fe	72.8 Gd (10)	Tb (10)
	31.6 Sn (50)	Tm(80)		71.9 Tb (10)	Pr (10)
	30.1 Yt (80)	Fe (5)		70.3 Eu (1000)	Mo(200)
	29.1 Cb (20)	Gd (20)	(300)Fe	69.6 Mo(15)	Ru (12)
	28.6 Cr (25)	Ho (12)		68.4 La (40)	Mo(30)
	27.6 V (200)	Tb (15)		67.3 Fe (30)	Mn(12)
	26.0 V (150)	La (100)Band		66.5 Ho (100)	La (8)
	25.3 Ni (30)	Ir (15)	(80)Ti	65.4 Fe (70)	Ce (35)
(150)Fe	24.5 V (100)	V (70)		64.9 Ce (40)	U (40)
	23.0 Pr (15)	Ce (8)	(100)Fe	63.2 Re (150)	Mo(10)
	22.4 Eu (200)	V (10)		62.7 Fe (15)	Pr (15)
	21.7 Sm (60)	Pr (15)		61.3 Sm (20)	V (15)
	20.5 Nd (200)	Ta (80)		60.6 Gd (20)	Ho (20)
	19.4 Pd (50)	Mo(15) Nd (10)		59.7 Ru (60)	Ce (15)
	18.8 Eu (125)	Mo(20)		58.7 V (18)	Th (10)
	17.9 Gd (20)	Ta (15)		57.9 Al (15)	Al (15)
	16.1 Gd (10)	W (10)		56.4 Yb (1500)	Ce (35)
(400)Fe	15.6 Th (10)	Ce (6)		55.8 Er (8)	W (6)
	14.7 Ni (15)	Ce (20)	(100)Fe	54.8 Tb (25)	Tb (25)
	13.0 Mo(20)	Ho (12)		53.1 Ho (30)	Ti (15)
	12.2 Re (20)	W (5)		52.3 Bi (500)	Hf (40)
	11.8 Er (12)	Nd (5)		51.3 Cb (30)	Mn(10)
	10.9 Mo(30)	U (30)		50.3 Sm (125)	Hf (10)
	09.2 Mo(12)	Ce (6)		49.9 Fe (8)	Cb (5)
	08.6 Mo(20)	Rh (10)		48.9 Sm (80)	Ta (60)
	07.2 Re (10)	Tb (10)		47.4 Eu (1000)	Pd (50)
	06.3 Yt (12)	Ru (6)		46.4 Fe (40)	Tb (15)
	05.8 Eu (40)	Gd (10)		45.9 Ag (30)	Co (25)
	04.9 V (60)	Sm (30)		44.5 Rh (50)	Mo(20)
	03.1 Sm (100)	Cb (15)		43.3 Sr (30)	Fe (25)
(45)Fe	02.5 La (300)	Band		42.7 Pd (100)	Eu (80)
	01.3 Ce (50)	Ca (15)		41.2 La (20)	Ti (15)
	00.8 Sm (200)	La (100)Band		40.0 Sr (20)	Ru (12)
	5500			39.0 Yb (200)	Fe (30)
	99.4 Rh (300)	Eu (40)	(50)Fe	38.5 Gd (20)	Ti (10)
(35)Ca	98.7 Ta (60)	Co (50)		37.0 Sm (50)	Mn(40)
	97.9 C (50)	Band		36.1 Eu (30)	Tb (25)
	96.3 Mo(10)	Sn (5)	(1000)Ba	35.5 Rh (80)	Fe (50)
	95.8 Ce (25)	Nd (8)		34.6 Fe (20)	Sr (20)
(35)Ca	94.4 Nd (150)	Co (40)		33.0 Mo(200)	Eu (30)
	93.7 Ni (40)	Ba (12)		32.6 Re (100)	Na (15)
	92.2 Ni (150)	V (50)		31.1 Pr (20)	W (18)
	91.8 Gd (20)	Mo(20)		30.7 Co (500)	Ti (30)
	90.7 Co (500)	Ca (15)		29.4 Pd (10)	Nd (10)
	89.3 Ni (20)	Ti (10)		28.4 Mg(60)	Nd (20)

Wave-length Table-chart

	27.5 Yt (100)	Tl (30)		66.4 Yt (150)	Sm (80)
	26.8 Sc (100)	Eu (60)		65.4 Ag (1000)	Mo (20)
	25.5 Fe (40)	Tb (25)		64.3 La (25)	Tb (10)
	24.3 Hf (40)	Tb (40)	(100)Fe	63.2 Cr (15)	Hf (10)
	23.2 Co (300)	Os (100)	(50)Fe	62.9 Ni (20)	Er (20)
	22.9 Ce (100)	Ce (15)		61.2 Ta (80)	Th (12)
	21.8 Sr (50)	Re (20)		60.5 Ti (30)	Re (30)
	20.4 Sc (80)	Mo (20)		59.8 Tb (25)	U (8)
(200)Ba	19.1 Tb (10)	W (7)		58.1 V (15)	Ta (12)
	18.9 Ta (100)	Er (12)		57.4 Mn (25)	Os (25)
	17.3 La (30)	Ce (10)		56.1 Ru (40)	Er (30)
	16.1 Sm (200)	Mn (50)	(300)Fe	55.6 La (200)	Fe (50)
	15.2 La (15)	Gd (6)		54.5 Co (300)	Ru (100)
(150)Ti	14.5 Fe (50)	W (50)		53.0 Sm (100)	Os (18)
	13.5 Pr (10)	Er (8)		52.9 Eu (1000)	Co (25)
(125)Ti	12.5 Sm (80)	Ce (50)		51.5 Eu (1000)	Nd (25)
	11.5 U (30)	V (25)		50.8 Sr (30)	Mo (30)
	10.5 Eu (300)	La (200) Ru (100)		49.5 Ir (35)	Ce (25)
	09.1 Pr (50)	Sm (40)		48.9 Ti (12)	Er (12)
	08.2 Mo (20)	Nd (15)		47.9 Re (20)	Nd (12)
	07.7 V (60)	Nd (8)	(300)Fe	46.9 Ti (15)	Sc (15)
(150)Fe	06.4 Mo (200)	La (50)	(150)Fe	45.0 Rh (25)	Ce (10)
	05.8 Mn (40)	Yb (40)		44.5 Co (400)	Hf (20)
	04.1 Sr (60)	Cb (30)		43.5 Eu (125)	Os (50)
(60)Ti	03.8 La (100)	W (45)		42.2 Nd (25)	Cr (18)
	02.6 La (20)	Er (8)		41.9 Tb (15)	Rh (15)
(150)Fe	01.3 La (200)	Mo (40)		40.4 Ti (12)	W (10)
	00.8 Eu (20)	Ta (20)		39.5 Rh (20)	Mo (12)
	5400			38.2 Yt (20)	Tb (15)
	99.4 Ta (60)	Th (12)		37.2 Cb (30)	Mo (30)
	98.2 Sm (80)	Ho (20)		36.9 Co (25)	Ti (10)
(150)Fe	97.5 Yt (20)	Pr (10)		35.2 Ta (80)	Ni (50)
	96.6 Ru (15)	U (12)	(300)Fe	34.5 V (50)	Er (12)
	95.1 Eu (125)	Co (15)		33.8 Er (12)	Ce (8)
	94.7 Ta (50)	Nd (15)		32.5 Mn (40)	Er (12)
	93.7 Sm (80)	Mo (20)		31.5 Rb (100)	Ta (60)
	92.9 U (60)	W (50)		30.2 Ce (10)	Sm (4)
	91.7 Er (12)	Ce (10)	(500)Fe	29.6 Ti (25)	Er (12)
(70)Ti	90.1 Ta (60)	Mo (20)		28.2 Tb (10)	W (4)
	89.6 Co (150)	V (18)		27.5 Ru (25)	Er (12)
	88.6 Eu (500)	Ti (30)		26.9 Eu (200)	Mo (20)
(50)Fe	87.1 V (20)	W (15)		25.4 Rh (25)	Th (15)
	86.1 Sr (40)	W (20)	(400)Fe	24.0 Ba (100)	Rh (100)
	85.4 Sm (40)	Er (30)		23.2 Rh (15)	Tb (15)
	84.3 Ru (60)	Sc (60)		22.8 Er (30)	Cb (15)
	83.3 Co (500)	Co (150)		21.0 Eu (125)	Lu (50)
	82.2 La (25)	U (12)		20.3 Mn (60)	Sm (25)
	81.9 Sc (60)	Mn (50) U (30)		19.1 Ta (80)	Er (20)
	80.8 Sr (100)	Ru (25) U (15)		18.8 Ru (20)	V (15)
	79.4 Ru (40)	Ho (12)		17.5 Os (40)	Mo (20)
	78.4 Pt (50)	Nd (15)		16.0 Sm (100)	Os (80)
(70)Ti	77.7 Co (40)	Os (30)	(500)Fe	15.2 V (75)	Th (20)
(80)Fe	76.6 Lu (500)	Ni (400)		14.6 Er (50)	Ce (12)
	75.7 Pt (60)	Ta (40)		13.6 Mn (30)	Pr (15)
(100)Fe	74.9 Ti (30)	Ti (12)		12.1 Os (15)	Gd (8)
(100)Fe	73.9 Mo (50)	Ba (10)		11.8 Eu (80)	Ni (40)
	72.3 Eu (1000)	Ti (12)	(200)Fe	10.9 Ta (15)	Tb (10)
	71.5 Ag (500)	Ti (25)	(50)Ti	09.7 Cr (300)	Ce (50)
	70.4 Co (50)	Mn (50)		08.1 Co (30)	Ta (20)
	69.3 Co (125)	Os (30)		07.5 Co (100)	Mn (60)
	68.4 Ho (20)	Ce (15)		06.3 Mo (15)	W (8)
	67.0 Eu (10)	V (10)	(400)Fe	05.7 Sm (80)	Eu (40)

(300)Fe	04.1 Ta (80)	Rh (50)			43.3 Co (600)	Er (30)
	03.8 Fe (30)	Yb (20)			42.7 Co (800)	Ta (80)
	02.7 Eu (1000)	Lu (150)		(200)Fe	41.3 Co (300)	Mn(200) Ta (150)
	01.0 Ru (125)	Co (100) V (100)			40.6 La (80)	Cr (50)
(150)Fe	00.5 Cr (30)	Mo(20)		(200)Fe	39.9 Co (100)	K (40)
	5300				38.6 Re (10)	Ti (8)
	99.4 Mn(40)	Ce (12)			37.5 Gd (20)	W (15)
(70)Fe	98.2 Sm (15)	Th (8)			36.1 Co (50)	Ta (40)
(400)Fe	97.1 Ti (60)	Ta (30)			35.1 Yb (150)	Ru (100)
	96.4 Re (15)	Ti (12)			34.8 Co (70)	Ru (60)
	95.9 Ta (80)	Pd (50)			33.6 Co (100)	Re (30)
	94.6 Mn(50)	Mo(20)			32.6 Co (200)	Ru (40)
(150)Fe	93.1 V (100)	Ce (30)			31.4 Co (500)	Re (80)
	92.9 Eu (150)	Sc (20)			30.5 Ce (25)	Er (12)
	91.4 Fe (25)	Cr (15)			29.8 Sr (40)	Rh (30)
	90.4 Rh (125)	Pt (50)		(400)Fe	28.0 Fe (150)	Ta (40)
(60)Fe	89.3 Ta (100)	Ti (30)			27.4 Re (100)	Mo(20)
	88.5 Ta (40)	W (15)			26.9 Th (10)	Co (10)
	87.9 Sm (50)	Cr (18)			25.2 Co (300)	Co (25)
	86.9 Cr (20)	Ce (15)		(400)Fe	24.1 Sm (25)	Hf (20)
	85.1 V (45)	Ru (25)			23.2 K (40)	Sm (10)
	84.1 Gd (10)	Er (8)			22.0 Fe (30)	Pr (30)
(400)Fe	83.3 V (40)	Sm (3)			21.8 Sm (50)	Re (40)
	82.9 Th (12)	Er (12)			20.5 Sm (100)	Th (8)
	81.7 Co (300)	Rh (100)			19.8 Nd (60)	Tb (40)
	80.9 La (300)	Sm (10)			18.6 Cb (100)	Cr (30)
	79.0 Rh (100)	Fe (35)			17.2 Re (40)	Mo(12)
	78.7 Tb (10)	Fe (5)			16.7 Co (300)	Eu (15)
	77.0 Re (300)	Mn(40)			15.0 Mo(20)	Mo(10)
	76.9 Eu (200)	Os (50)			14.7 Rh (40)	F (50) Band
	75.9 Tb (40)	Cb (15)			13.8 Mo(25)	Ce (8)
	74.1 W (12)	W (10)			12.6 Co (400)	Sm (100)
	73.8 Hf (15)	Fe (15)			11.6 Hf (100)	U (18)
	72.4 Mo(20)	Gd (20)			10.3 F (80)	Band Co(20)
(700)Fe	71.4 Ni (30)	Nd (20)			09.2 Ru (125)	Tb (10)
	70.6 Gd (25)	Cr (10)			08.5 U (25)	Ba (10)
(150)Fe	69.5 Co (500)	Re (40)		(125)Fe	07.3 Tm(100)	Yb (40)
	68.3 Sm (80)	Pt (50)			06.2 Mo(15)	Gd (10)
(200)Fe	67.4 Er (12)	Mo(12)			05.5 Re (25)	Tm(20)
	66.7 Sm (10)	Ti (8)			04.4 F (100)	Band Ru (60)
	65.4 Fe (40)	Ta (12)			03.8 Eu (300)	La (100)
(200)Fe	64.8 Mo(70)	Sm (50)		(300)Fe	02.3 Ba (20)	F (100) Band
	63.6 Yb (25)	Ce (15)		(150)Pt	01.0 Co (700)	La (300)
	62.7 Co (500)	Rb (50)			00.7 Cr (25)	Tb (10)
	61.5 Eu (300)	Ru (100)			**5200**	
	60.8 Eu (150)	Mo(100)			99.8 Hf (8)	U (6)
	59.1 Co (300)	K (40)		(40)Ti	98.0 Hf (80)	F (100) Band
	58.9 Co (40)	Yb (15)		(70)Ti	97.2 Tb (10)	Th (8)
	57.6 Eu (1000)	La (40)			96.8 F (100)	Band
	56.7 Eu (40)	Sc (40)		(50)Ti	95.6 Pd (200)	Ta (40)
	55.0 Eu (200)	Mo(12)			94.5 Eu (300)	V (18)
	54.3 Rh (300)	C (100) Band			93.1 Nd (60)	Eu (50)
(60)Fe	53.4 Co (500)	Ce (50)			92.9 F (150)	Band Rh (80)
	52.0 Co (500)	Yb (100)			91.0 F (200)	Band Eu (200)
(50)Ti	51.6 Eu (150)	Yb (50)			90.8 La (60)	Fe (15)
	50.4 Tl (5000)	Cb (150)			89.2 Eu (125)	Tb (15)
(12)Ca	49.0 Ta (80)	Co (80)			88.5 Fe (30)	Ti (15)
	48.3 Cr (150)	W (30)		(100)Fe	87.2 Eu (125)	Cr (40)
	47.4 Co (80)	Yb (40)			86.8 Ce (12)	Er (12)
	46.4 Tm(40)	Hf (10)			85.7 Eu (80)	Pr (40)
	45.8 Cr (300)	Yb (20)			84.0 Ru (100)	Ti (18)
	44.1 Cb (400)	Er (30)		(400)Fe	83.6 Co (125)	Ti (50)

	82.8 Eu (1000)	Sm (100)			20.0 Cu (100)	Pr (80)
(300)Fe	81.7 Tb (25)	V (10)		(60)Ti	19.0 Cb (100)	Pr (50)
	80.6 Co (500)	Mo(50)			18.2 Cu (700)	Ta (80)
	79.8 Ta (60)	Er (30)		(150)Fe	17.3 Eu (125)	Gd (25)
	78.2 Re (100)	Eu (15)		(300)Fe	16.2 V (40)	Th (12)
	77.0 Yb (200)	Ba (10)		(200)Fe	15.0 Eu (1000)	Er (20)
	76.1 Co (400)	Cb (200)			14.1 Cr (18)	Tb (15)
	75.5 Re (500)	Eu (30)			13.3 Eu (150)	V (35)
	74.2 Ce (50)	Sm (10)			12.7 Co (300)	Ta (60)
(130)Fe	73.1 Nd (25)	Cr (15)			11.8 La (300)	Co (100)
	72.4 Eu (400)	Sm (50)		(200)Ti	10.3 Co (100)	Co (50)
(100)La	71.9 Eu (2000)	Cb (200)Sm (150)			09.0 Ag (1500)	Sm (20)
(400)Fe	70.3 Re (200)	Ca (20)		(200)Fe	08.4 Cr (500)	Pd (10)
(800)Fe	69.5 Rh (50)	W (12)			07.8 Ti (25)	Sm (25)
	68.5 Co (400)	Gd (20)			06.0 Cr (100)	Ti (40)
	67.0 Ba (25)	Tm(10)			05.7 Yt (50)	Sm (15)
(500)Fe	66.4 Eu (1000)	Co (600)		(125)Fe	04.5 Cr (400)	La (50)
(70)Ti	65.9 Eu (400)	Os (30)			03.2 W (30)	Cb (15)
(15)Ca	64.1 Cr (100)	Hf (50)		(300)Fe	02.3 Sm (50)	Os (30)
(300)Fe	63.3 Ti (30)	Eu (30)			01.0 Ti (30)	Ce (15) Pb (10)
	62.1 Tb (40)	Ca (20)			00.9 Eu (300)	Sm (200)
	61.8 Au (40)	Ca (20)			**5100**	
	60.9 V (30)	Hf (30)			99.8 Eu (500)	Ru (20)
	59.7 Pr (125)	Mo(60)		(80)Fe	98.7 Tb (15)	Th (15)
	58.3 Sc (12)	Th (6)			97.7 Gd (25)	Mn(10)
	57.6 Co (400)	Ru (25)			96.4 Cr (50)	Mn(30)
	56.9 Sr (90)	Eu (20)		(100)Fe	95.4 Ru (100)	V (40)
	55.3 Mn(50)	Ti (40)		(200)Fe	94.9 V (30)	Pr (20)
(50)Fe	54.6 Co (200)	Cr (18)			93.1 Rh (200)	Cb (100)
(70)Fe	53.4 La (100)	In (30)		(400)Fe	92.3 Ti (150)	V (100)Co (100)
	52.7 Sm (40)	Ti (35)		(400)Fe	91.4 Nd (40)	Ce (30)
	51.8 Sm (150)	Gd (30)			90.8 Th (10)	Tb (10)
(150)Fe	50.6 Fe (30)	Ti (12)			89.1 Cb (80)	Eu (10)
	49.9 Co (200)	Nd (60)		(80)Ti	88.7 Ca (50)	La (50)
	48.6 Eu (80)	Re (25)			87.4 Ce (50)	Hf (20)
(50)Fe	47.9 Co (500)	Cr (60)			86.9 Cb (50)	Ti (20)
	46.1 Er (20)	Gd (20)			85.9 Ti (8)	Rh (8)
	45.9 Ce (30)	Mo(25)			84.1 Rh (100)	Cr (60)
	44.1 Yb (50)	Ta (40)			83.6 Mg(500)	La (300)
	43.3 Cr (50)	Fe (20)			82.6 Nd (8)	Th (5)
(125)Fe	42.4 Mo(50)	W (25)			81.9 Zn (200)	Re (25)
	41.4 Cr (12)	U (6)			80.3 Cb (150)	Ce(15)
	40.8 Mo(80)	V (50)			79.7 Nd (25)	Gd (20)
	39.2 Eu (80)	C (70)Band			78.9 Re (100)	Sm (100)
(50)Ti	38.5 Sr (90)	Mo(80)			77.3 La (150)	Cr (50)
	37.1 Rh (100)	Mo(15)			76.0 Co (500)	Ni (70)
	36.1 Eu (25)	Re (20)			75.9 Rh (200)	Sm (60)
	35.2 Co (100)	Fe (35)			74.1 Mo(70)	Ce (25)
	34.2 La (200)	Pd (50)		(125)Ti	73.7 Pr (100)	La (20)
	33.8 Ti (35)	Eu (30)			72.6 Mg(200)	Sm (80)
(800)Fe	32.9 Cb (50)	Mo(20)		(300)Fe	71.5 Ru (150)	Mo(30)
	31.0 Mo(20)	Th (12)			70.6 Tb (20)	Hf (10)
	30.2 Co (300)	Ta (60)			69.6 Sm (50)	V (18)
(200)Fe	29.8 Sr (70)	Er (30)		((80)Ti	68.9 Ni (70)	Os (8)
	28.7 Sm (60)	Tb (40)		(700)Fe	67.4 Mg(100)	Mo(25)
(400)Fe	27.1 Pt (80)	Ti (10)		125)Fe	66.7 Eu (125)	Sm (125)
(200)Fe	26.8 Ti (30)	Cr (15)		(50)Fe	65.4 Co (30)	Sr (15)
(60)Fe	25.1 Sr (70)	V (40)		(70)Fe	64.3 Cb (150)	Er (30)
(90)Ti	24.9 Ti (70)	W (50)			63.8 Pd (300)	Ta (40)
(50)Ti	23.4 Eu (700)	Ru (20)		(300)Fe	62.2 Sm (15)	W (9)
	22.1 Sr (70)	Co (50)			61.8 Ta (80)	Pr (40)
	21.7 Cr (25)	Tb (15)			60.3 Cb (200)	Eu (200)

202 Manual of Spectroscopy

(50)Ba	59.9	V (40)	Fe (35)	(200)Fe	98.7	Gd (20)	Mo(20)
	58.6	Rh (80)	La (50)		97.5	Mo(40)	K (25)
	57.4	La (40)	Rh (25)	(35)Fe	96.5	Re (100)	Ni (50)
	56.3	Co (300)	Sr (80)		95.2	Cb (60)	Mo(20)
	55.5	Rh (150)	Ru (125)Sm (125)		94.9	Cb (100)	Ni (25)
	54.0	Co (200)	Sm (125)		93.8	Ru (60)	Er (8)
	53.6	Na (600)	Cu (600)		92.2	Gd (30)	Nd (20)
(90)Ti	52.6	Cb (100)	Pr (15)		91.8	Cr (25)	Ce (15)
(70)Fe	51.9	Ru (40)	Nd (5)	(40)Fe	90.6	Rh (150)	Ta (60)
(150)Fe	50.8	Mn(40)	Ta (15)		89.1	Tb (40)	Eu (20)
	49.0	Na (400)	Co (100)		88.9	Sm (30)	Sm (25)
	48.7	V (60)	Fe (35)	(70)Ti	87.0	Ta (60)	Yt (50)
(90)Ti	47.4	Ru (60)	Au (40)		86.9	Sc (60)	Ce (6)
	46.7	Co (400)	Ni (150)		85.8	Cd (1000)	Sc (80)
(100)Ti	45.4	La (100)	Co (80)		84.0	Ni (300)	K (20)
	44.6	Cr (30)	Er (12)	(200)Fe	83.3	Sc (100)	Ce (12)
	43.6	Ta (30)	Er (12)		82.3	Ni (100)	Ta (30)
(125)Fe	42.9	Fe (100)	Ni (100)		81.1	Ni (150)	Sc (100)
(100)Fe	41.7	Ta (40)	Gd (20)		80.5	Ni (200)	Mo(20)
	40.8	Gd (25)	Ce (10)	(100)Fe	79.7	Fe (100)	Hf (40)
(200)Fe	39.4	Fe (125)	Ni (50)		78.9	Ca (300)	Sm (25)
	38.4	V (50)	W (20)		77.6	Er (12)	Cb (8)
(200)Fe	37.3	Ni (150)	Ce (22)		76.7	Yb (50)	Ta (50)
	36.5	Ru (125)	Ta (60)		75.8	Sc (12)	Pr (12)
	35.0	Lu (200)	Pr (50)	(80)Fe	74.3	Yb (200)	Mn(15)
	34.7	Cb (200)	Lu (20)		73.7	Tb (15)	Zr (6)
(200)Fe	33.6	Eu (150)	Pr (60)		72.9	Cr (35)	Ru (25)
	32.2	Gd (20)	Ta (15)	(40)Ti	71.1	Sm (100)	W (40)
(125)Fe	31.4	Tb (15)	Th (10)		70.2	Sc (20)	Tb (20)
	30.5	Nd (40)	Gd (20)	(40)Ti	69.4	Sm (150)	W (50)
	29.0	Eu (200)	Pr (100)	(400)Fe	68.7	Cr (20)	Ti (8)
	28.5	V (75)	Hf (10)		67.8	Ta (60)	Cr (50)
(100)Fe	27.3	Er (25)	Ru (20)		66.3	Sm (15)	Sm (10)
	26.1	Co (200)	Re (50)	(50)Ti	65.2	Cb (80)	Fe (40)
(100)Fe	25.6	Co (100)	Ni (50)	(150)Ti	64.6	V (50)	Au (40)
	24.7	Eu (60)	Sm (50)		63.9	Ce (20)	Eu (20)
(200)Fe	23.7	Nd (30)	Gd (20)		62.1	Ti (40)	Mo(20)
	22.9	La (150)	Co (150)		61.2	Th (12)	Th (4)
	21.5	Ni (20)	Tb (15)		60.9	Sm (30)	Tm(30)
(100)Ti	20.4	Cb (50)	Re (40)		59.4	Pt (60)	Mo(40)
	19.6	Er (8)	Yt (7)		58.0	Cb (50)	Re (40)
	18.3	Tb (15)	W (7)		57.7	Sm (100)	Ru (160)
	17.1	Sm (80)	Pd (50)		56.4	La (80)	Nd (15)
	16.6	Sm (100)	Mo(12)		55.5	W (20)	Mo(20)
	15.8	Ta (40)	Ni (80)		54.6	W (25)	Ba (12)
	14.5	La (150)	Eu (150)		53.2	W (60)	Pr (30)
(80)Ti	13.2	Co (100)	Cr (25)	(50)Ti	52.7	Sm (150)	W (10)
	12.2	Sm (125)	K (30)	(200)Fe	51.6	Cr (50)	Ni (50)
	11.9	Cu (15)	Sm (12)		50.5	La (80)	Gd (50)
(300)Fe	10.4	Pd (100)	Pr (160)	(400)Fe	49.8	Th (30)	Tb (15)
	09.7	Ta (30)	Mo(25)	(50)Fe	48.8	Ni (80)	Cr (30)
	08.8	Co (200)	Gd (30)		47.7	Mo(25)	Hf (15)
(100)Fe	07.4	Ru (40)	Tm(30)		46.8	La (80)	Ca (7)
	06.2	La (100)	Fe (25)		45.5	Pr (40)	Ti (25)
	05.5	Cu (500)	V (40)	(25)Fe	44.2	Sm (150)	Pt (60)
	04.4	Sm (125)	Re (50)		43.3	Ta (60)	Ti (30)
	03.0	Sm (150)	Os (30)		42.1	Ni (80)	Er (60)
(80)Fe	02.2	Ni (40)	Cb (20)	(425)Fe	41.7	Ca (30)	Ni (30)
	01.3	Ru (20)	Sc (12)		40.8	Hf (100)	Ti (40)
	00.1	Cb (100)	Sm (60)	(125)Ti	39.0	Cb (200)	Fe (100)
	5000			(100)Ti	38.4	Ni (50)	Mo(10)
	99.9	**Ni (230)**	**Sc (100)**		37.3	Ta (60)	Ta (20)

Wave-length Table-chart

```
(125)Ti  36.4 Sm (50)   Ce (10)              (80)Ti  75.9 Sm (80)   Cb (30)
(125)Ti  35.3 Ni (300)  Ni (70)                      74.1 Ru (20)   Pr (15)
         34.2 Tm(100)   Pr (30)             (100)Fe  73.1 Sm (40)   Ti (35)
         33.5 Eu (60)   Pt (30)                      72.1 Sm (80)   Gd (25)
         32.7 Ni (6)    Eu (6)                       71.9 Li (500)  Co (150)Ni (100)
         31.0 Sc (50)   Gd (30)                      70.3 La (125)  Fe (20)
         30.7 Mo(20)    Mn(8)               (50)Fe   69.1 Gd (100)  Ta (40)
         29.4 Eu (500)  Eu (30)                      68.5 Ta (60)   Ti (40)
(100)Fe  28.4 Sm (200)  Th (40)                      67.7 Cb (150)  Sr (20)
(120)Fe  27.2 Eu (40)   Re (25)             (300)Fe  66.0 Co (100)  Yb (30)
         26.9 Pr (80)   Cb (50)                      65.3 Cb (100)  Gd (100)
(100)Ti  25.5 Tb (15)   W  (8)                       64.5 Sm (80)   Ti (25)
(100)Ti  24.8 Tb (25)   Tb (15)                      63.7 Rh (100)  Nd (15)
         23.5 Sm (60)   Gd (20)                      62.2 Sr (40)   Al (15)
(150)Fe  22.2 Ti (100)  Eu (125)                     61.9 Sm (100)  Gd (100)
         21.9 Cr (20)   Ce (20)                      60.1 Eu (20)   Pr (10)
(100)Ti  20.0 Gd (20)   Th (10)                      59.1 Nd (35)   Ru (15)
         19.7 Pr (50)   Cr (18)                      58.7 Gd (125)  Ta (8)
 (80)Fe  18.4 Ni (70)   Pr (50)             (300)Fe  57.6 Fe (100)  Dy (20)
         17.5 Ni (100)  Cb (80)  Th (50)             56.6 Pr (40)   Re (30)
(100)Ti  16.1 Sm (80)   Mo(20)                       55.2 Ru (25)   U  (8)
         15.0 Gd (100)  W  (40)  Th (10)             54.8 Cr (100)  Nd (50)
(500)Fe  14.9 Ti (140)  V  (125)                     53.2 Ni (150)  Co (50)
 (80)Ti  13.1 Eu (125)  Cr (60)                      52.3 Sm (125)  La (50)
(300)Fe  12.0 Ni (70)   Ta (60)                      51.3 Pr (150)  Er (15)
         11.2 Ru (25)   Ce (12)                      50.8 F  (100)  Band   Mo(80)
 (50)Fe  10.8 Ni (50)   Gd (50)                      49.7 La (200)  Re (15)
 (50)Ti  09.6 Tm(50)    Ce (30)                      48.6 Sm (125)  Ce (18)
         08.2 Eu (30)   Er (12)                      47.5 Th (10)   Ti (7)
(200)Ti  07.2 Fe (25)   W  (15)                      46.7 Re (100)  La (100)
(300)Fe  06.1 W  (40)   Yt (8)                       45.4 Ni (90)   Cb (15)
(200)Fe  05.7 Tb (20)   Pb (20)                      44.8 Nd (50)   Cr (35)
         04.9 Mn(20)    Tb (15)                      43.9 Nd (10)   Mo(6)
         03.7 Ni (20)   Tb (10)                      42.4 Cr (125)  Gd (100)
         02.3 V  (90)   Fe (20)                      41.6 Mo(40)    Ti (30)
(300)Fe  01.8 Lu (100)  Ti (80)                      40.2 Pr (50)   Sm (30)
         00.3 Ni (150)  Cb (30)             (150)Fe  39.6 Pr (100)  Gd (30)
              4900                           (400)Fe  38.8 Eu (250)  Gd (150)Sm (125)
(200)Ti  99.4 La (400)  Mo(50)               (30)Ti  37.3 Ni (400)  Ta (40)
         98.2 Ni (150)  Gd (25)                      36.3 Cr (200)  Ta (100)
 (50)Ti  97.1 Fe (20)   Cb (15)                      35.5 Yb (200)  Ni (150)
         96.8 Ni (80)   La (25)             (400)Ba  34.0 La (150)  Fe (40)
         95.3 Mo(20)    Ti (12)              (50)Fe  33.3 Mo(30)    Sm (25)
(200)Fe  94.1 Lu (250)  W  (30)                      32.0 V  (15)   W  (12)
         93.8 La (80)   Gd (10)                      31.5 W  (30)   Mo(20)
         92.0 Sm (80)   Ru (25)                      30.7 Gd (80)   Fe (25)
(200)Ti  91.0 La (100)  Fe (80)                      29.5 Sm (40)   Th (10)
         90.6 Ce (10)   Os (6)              (100)Ti  28.2 Co (200)  U  (20)
(100)Ti  89.1 Sm (60)   Pr (50)              (50)Fe  27.4 Fe (20)   Th (5)
(100)Fe  88.0 Co (500)  Cb (150)                     26.0 Ta (60)   Mo(50)
         87.9 Mo(15)    Ru (15)             (1000)Fe 25.2 Ni (100)  V  (25)
         86.8 La (150)  W  (40)             (100)Fe  24.7 Pr (80)   Nd (80)
(100)Fe  85.5 Fe (100)  Re (40)                      23.9 Re (150)  Ta (60)
         84.1 Ni (500)  W  (30)                      22.2 Cr (200)  Sm (30)
(200)Fe  83.8 Fe (100)  W  (20)             (100)Ti  21.7 La (500)  Ta (50)
(200)Fe  82.8 Na (200)  W  (40)              (500)Fe 20.9 La (500)  Ta (150)Sm (125)
(300)Ti  81.7 Sm (50)   Re (15)              (80)Ti  19.8 Th (50)   Pd (12)
         80.1 Ni (500)  Ru (50)             (300)Fe  18.9 Ni (200)  Sm (125)
         79.1 Mo(100)   Co (60)                      17.4 Sm (15)   Mn(12)
 (80)Fe  78.6 Ti (70)   Na (15)                      16.6 Gd (30)   W  (20)
         77.6 Mo(50)    Rh (25)                      15.8 Gd (40)   Ti (30)
         76.3 Ni (40)   Ru (40)                      14.0 Pr (60)   Ta (50)
```

(125)Ti	13.9 Ni (200)	Sm (150)		52.5 Ni (150)	Ta (80)
	12.0 Ni (100)	Os (80)		51.6 Rh (80)	V (40)
	11.4 Eu (30)	Ta (30)		50.5 La (45)	Th (12)
(130)Fe	10.4 Sm (150)	W (30)		49.2 Os (18)	Nd (15)
(50)Fe	09.3 Mo(30)	Ti (12)	(60)Ti	48.3 Cb (100)	Pr (125)Sm (100)
	08.5 Re (20)	W (6)		47.7 Sm (150)	Ce (12)
	07.7 Ta (50)	Mo(30)		46.4 Ta (100)	Cr (18)
	06.9 Pr (50)	Er (18)		45.6 Er (50)	Yt (30)
	05.0 Cr (30)	W (12)		44.2 Sm (150)	Mn(80)
	04.4 Ni (400)	Ta (80)		43.4 Co (300)	Rh (100)W (50)
(500)Fe	03.3 Cr (125)	Mo(80)		42.4 Rh (50)	U (12)
	02.8 Ba (15)	W (15)		41.7 Sm (100)	Pr (5)
	01.8 La (50)	Sm (40)	(125)Ti	40.2 Co (700)	Mn(50)
	00.7 Sm (100)	Er (30)		39.6 Lu (50)	La (40)
	4800			38.6 Ni (150)	Mn(50)
(150)Ti	99.5 Co (400)	La (400)		37.6 Sm (100)	Pr (80)
	98.4 Th (12)	Th (10)		36.8 Cr (80)	Sm (50)
	97.1 Co (15)	Ce (10)		35.2 Gd (100)	Tm(20)
	96.9 Nd (60)	Pr (25)		34.2 Gd (125)	Sm (100)
	95.5 Ru (12)	Ru (10)		33.3 Sm (80)	Mo(25)
	94.3 Gd (200)	Sm (60)		32.0 Sr (200)	Pr (100)Ta (100)
	93.3 Sm (150)	Ti (10)		31.1 Ni (200)	Tm(50)
	92.4 W (25)	Sr (15)		30.5 Mo(125)	Eu (20)
(70)Fe	91.4 Sr (40)	Sm (40)		29.0 Ni (300)	Cr (200)Sm (200)
(100)Fe	90.7 Nd (30)	Cb (15)		28.4 Mo(25)	Pr (15)
	89.1 Re (2000)Ce (20)			27.4 V (20)	Pr (20)
	88.5 Cr (100)	W (20)		26.6 Pr (40)	Os (18)
	87.0 Cr (125)	Mo(20)		25.4 Ta (150)	Nd (100)
	86.9 W (50)	Ni (30)		24.0 La (150)	Zr (20)
(150)Ti	85.0 Cr (60)	U (18)		23.5 Mn(400)	Cr (25)
	84.9 Cr (25)	Eu (15)		22.9 Pr (125)	Ce (25)
	83.9 Ta (150)	Gd (100)	(200)Fe	21.0 Gd (100)	Ni (25)
	82.4 Ce (30)	Ti (25)	(125)Ti	20.4 Nd (40)	Er (25)
	81.9 Gd (100)	Gd (60)	(40)Ti	19.5 Ta (100)	Mo(80)
	80.0 Cr (25)	V (20)		18.2 Yt (30)	Band
	79.1 Pr (30)	Ta (25)		17.5 Pd (40)	Mo(25)
(100)Ca	78.1 Fe (80)	W (30)		16.8 Gd (50)	Cb (50)
	77.8 Pr (50)	Ba (30)		15.8 Sm (125)	Os (60)
	76.3 Sr (200)	Pr (20)		14.2 Cr (100)	Pr (30)
	75.9 Gd (100)	V (40)		13.4 Co (1000)Co (100)	
	74.1 Ag (30)	Ni (25)		12.7 Ta (150)	Ti (18)
	73.4 Ni (200)	Gd (200)		11.3 Nd (60)	Au (50)
(100)Fe	72.1 Sr (25)	Er (25)		10.5 Zn (400)	Cb (100)
(200)Fe	71.3 Gd (100)	Ta (50)		09.0 La (150)	Cr (12)
(100)Ti	70.0 Gd (200)	Cr (150)Ni (100)		08.8 Ni (25)	Mo(25)
	69.9 Sm (125)	Ru (125)		07.4 Gd (100)	V (40)
(100)Ti	68.2 Mo(50)	Sr (20)		06.9 Ni (150)	Cr (80)
	67.8 Co (800)	Eu (30)	(70)Ti	05.8 Gd (200)	Ce (18)
	66.2 Ni (300)	Gd (100)		04.0 La (150)	Cr (35)
	65.0 Gd (400)	Os (80)		03.5 Gd (100)	Sm (10)
	64.7 V (30)	Ta (20)		02.5 Gd (100)	Ru (7)
	63.9 Ni (30)	Th (20)		01.0 Gd (200)	Cr (200)
	62.6 Gd (100)	Mn(40)		00.4 Hf (50)	Gd (100)
	61.8 Cr (125)	Gd (100)		**4700**	
	60.9 La (100)	Mo(25)	(80)Ti	99.9 Cd (300)	W (50)
(150)Fe	59.7 Sm (80)	Nd (60)		98.8 Sm (50)	Pr (25)
	58.2 Mo(20)	U (15) Th (10)		97.1 Nd (30)	Mn(25)
	57.3 Ni (100)	Cr (25)		96.1 Cr (125)	Co (100)
(100)Ti	56.0 Gd (80)	Gd (80)		95.8 Co (100)	Ru (20)
	55.4 Ni (400)	Cr (20)		94.3 Ru (25)	Mo(12)
	54.3 Sm (125)	Yt (100)		93.9 Os (300)	Mo(30)
	53.6 Pr (20)	Pt (15)	(70)Ti	92.8 Co (600)	Au (200)Cr (200)

Wave-length Table-chart

0200)Fe	91.4 Re (200)	Gd (150)Sm (150)		29.2 Sc (100)	Cr (30)	
	90.3 Cr (100)	Hf (20)	(150)Ir	28.4 La (400)	Gd (150)Sm (150)	
(100)Fe	89.3 Cr (300)	Sm (50)		27.9 Co (300)	Mn(150)	
	88.1 Pd (200)	Fe (40)	(80)Ba	26.0 Sm (100)	Yb (45)	
	87.9 W (15)	Cr (12)		25.1 W (40)	Re (20)	
(150)Fe	86.5 Ni (300)	Yb (50)		24.4 Cr (125)	Tm(35)	
	85.8 Sm (100)	Lu (100)		23.1 Cr (125)	Gd (100)	
	84.6 Gd (100)	Zr (40)		22.5 Bi (1000)Zn (400)Ta (200)		
	83.4 Mn(400)	Sm (150)Pr (125)		21.5 V (15)	Gd (20)	
	82.7 Hf (40)	Ce (10)		20.1 Sm (20)	Ru (15)	
	81.4 Co (400)	Gd (200)		19.8 La (200)	Sm (125)	
	80.0 Co (500)	Ta (50)		18.4 Cr (200)	Sm (100)	
	79.3 Sc (80)	Nd (30)		17.7 Sm (100)	Sm (80)	
	78.2 Co (100)	Ir (50)		16.1 Sm (150)	La (100)	
	77.8 Sm (100)	Eu (40)		15.7 Ni (200)	Sm (100)	
	76.3 Co (300)	Mo (40)	(50)Fe	14.4 Ni (1000)Sm (50)		
	75.1 Cr (35)	Sm (30)		13.6 Eu (400)	Sm (100)	
	74.1 Sm (100)	Mn(50)		12.9 La (100)	Ni (30)	
	73.9 W (30)	Hf (25)		11.9 Gd (30)	Zr (15)	
	72.3 Cr (100)	W (12)	(100)Ti	10.1 Zr (60)	Sm (50)	
	71.1 Co (500)	Cr (18)		09.7 Mn(150)	Ru (150)Gd (100)	
	70.1 Sm (40)	Cr (35)	(50)Fe	08.0 Cr (200)	Cb (50)	
	69.3 Ru (20)	Ti (12)	(100)Fe	07.2 Mo(125)	Pr (80)	
	68.0 Co (300)	Ta (150)		06.0 Ta (200)	Nd (50)	
	67.8 Cr (100)	Gd (100)Co (100)		05.0 Re (40)	Gd (25)	
	66.8 La (100)	Mn(80)		04.5 Cu (200)	Sm (200)	
	65.2 Pr (80)	Mn(60)		03.8 Ni (200)	La (200)Mg (8)	
	64.2 Cr (200)	Mo (50)		02.4 Tb (80)	Gd (50)	
	63.9 Ni (150)	Gd (25)		01.5 Ni (250)	Ta (150)Mn(100)	
	62.6 Ni (150)	Mn(100)	(25)Ba	00.6 Cr (50)	W (50)	
	61.5 Mn(60)	Pr (25)		**4600**		
	60.2 Sm (150)	Mo(125)		99.6 La (200)	Sm (50)	
(100)Ti	59.2 Tm(25)	Nd (20)	(100)Ti	98.3 Co (300)	Eu (300)	
(125)Ti	58.1 Eu (60)	Mo (40)		97.4 Gd (100)	Cu (60)	
	57.8 Ru (125)	Pr (100)		96.4 Nd (30)	Rh (30)	
(100)Co	56.1 Cr (300)	Ni (250)Ta (150)		95.7 Pr (60)	Cr (50)	
	55.3 Sm (100)	Cr (70)		94.3 Gd (50)	Th (15)	
	54.0 Mn(400)	Co (200)Ni (100)		93.2 Co (500)	Ta (150)	
	53.1 Sc (80)	V (30)		92.5 La (200)	Os (80)	
	52.4 Ni (150)	Tb (100)Cr (100)	(100)Ba	91.9 Ta (400)	Ti (125)Fe (80)	
	51.8 Na (20)	Er (12)		90.3 Nd (30)	Ru (25)	
	50.7 Sm (60)	La (40)		89.3 Cr (80)	U (30)	
	49.6 Co (500)	Cb (100)		88.2 Eu (100)	Zr (50)	
	48.7 La (100)	Re (50)		87.8 Zr (125)	Sm (100)	
	47.4 Fe (30)	Ce (80)		86.2 Ni (200)	Cr (20)	
	46.9 Pr (100)	Co (100)	(25)Ca	85.2 Ta (80)	Eu (60) Ge (20)	
	45.6 Sm (250)	Cr (80)		84.9 Pr (125)	Ta (100)Ru (100)	
	44.9 Pr (100)	Pr (80)		83.3 Gd (100)	Nd (50)	
	43.6 Gd (300)	La (300)Sc (100)		82.3 Co (500)	Yt (60)	
(100)Ti	42.7 V (20)	Ti (15)	(200)Ti	81.8 Ta (200)	Ru (100)	
	41.0 Sc (100)	Sm (80)		80.1 Zn (300)	W (150)	
	40.5 Eu (500)	La (150)Ta (100)		79.1 Gd (25)	Re (20)	
	39.1 Mn(150)	Zr (100)	(150)Fe	78.1 Cd (200)	Ta (40)	
	38.1 Gd (50)	Hf (20)		77.8 Tm(50)	W (25)	
	37.3 Cr (200)	Co (150)Sc (100)		76.9 Sm (100)	Tb (25)	
(125)Fe	36.7 Pr (125)	Eu (60)	(50)Ti	75.0 Rh (100)	Cb (50)	
	35.7 Gd (150)	Fe (10)		74.7 Cu (200)	Sm (80) Yt (80)	
(100)Sc	34.8 Co (150)	Gd (100)Pr (100)	(40)Ba	73.6 Fe (20)	Dy (10)	
	33.3 Tm(80)	Ru (40)		72.0 Cb (150)	Pr (100)	
	32.6 Gd (300)	Ni (100)		71.8 La (100)	Mn(100)	
(50)Ti	31.8 Ni (100)	Mo(100)Dy (30)		70.4 Sc (100)	V (60)	
	30.7 Cr (100)	Ta (100)		69.1 Ta (300)	Cr (50)	

(125)Fe	68.9 La (200)	Na (200)Ag (200)			06.2 Ni (100)	Cb (50)
(150)Ti	67.5 Fe (150)	Ni (160)			05.3 Mn(150)	La (100)
	66.9 Ni (50)	Cr (50)			04.9 Ni (300)	Ta (200)
	65.0 Eu (30)	Cr (20)			03.4 Tm(35)	U (25)
	64.6 Pr (100)	Na (80) Cr (70)		(300)Fe	02.8 Li (800)	Ta (100)
	63.4 Co (700)	Os (100)La (100)			01.4 Ta (60)	Gd (40)
	62.5 La (150)	Mo(40)			00.3 Ni (200)	Cr (150)
	61.1 Ta (300)	Eu (180)			**4500**	
	60.3 Eu (50)	Pr (25)		(50)Ba	99.0 Ru (100)	Tm(80)
	59.8 W (200)	Lu (10)		(50)Fe	98.1 Yb (25)	Cr (20)
	58.0 Lu (100)	Tb (40)			97.1 Os (100)	Eu (40)
	57.3 Co (100)	W (50)			96.9 Co (400)	Tm(25)
(150)Ti	56.4 Ir (60)	Cr (50)			95.3 Sm (100)	Os (80)
	55.4 La (150)	Hf (50)			94.0 Eu (500)	Co (400)
	54.3 Ru (135)	Cr (70)			93.1 Cs (1000)Sm (50)	
	53.4 Eu (25)	Tm(25)		(200)Fe	92.6 Ni (200)	Ru (100)
	52.1 Cr (200)	Re (30)			91.3 Cr (200)	Sm (100)
	51.1 Cu (250)	Pr (125)Cr (100)			90.8 Yb (40)	Mo(15)
(60)Ti	50.0 Nd (25)	Yb (15)			89.3 Dy (70)	Ti (40)
	49.4 Cr (60)	Nd (40)			88.6 Co (100)	Mo(25)
	48.6 Ni (400)	Cb (50)			87.8 Cr (30)	Ir (20)
(125)Fe	47.6 Ru (125)	Mo(25)			86.9 Cu (250)	V (40)
	46.1 Cr (100)	Nd (60)		(125)Ca	85.8 Cr (10)	Ir (8)
(100)Ti	45.1 Ru (100)	Tb (60)			84.4 Ru (150)	Sm (60)
	44.3 Co (70)	Eu (50)		(150)Fe	83.8 Ta (150)	Gd (40)
	43.5 Pr (60)	Er (50) Yt (50)			82.3 Yb (80)	Mn(20)
	42.2 Sm (100)	Mn(50)		(100)Ca	81.5 Co (1000)Mn(125)	
	41.1 Nd (80)	Tb (40)			80.1 Co (300)	Cr (300)Ta (200)
	40.6 Zr (150)	Band		(75)Ba	79.6 Mn(50)	Nd (25)
(60)Ti	39.9 Ti (80)	Pr (60)		(80)Ca	78.5 Pr (40)	Tb (70)
(80)Fe	38.0 Os (25)	Nd (15)			77.6 Sm (100)	V (40)
(100)Fe	37.5 Ti (20)	Zr (100)Band			76.2 Yb (90)	Pr (50)
(20)Ba	36.6 Gd (25)	Tb (20)			75.5 Zr (50)	Mn(50)
	35.6 Ru (125)	Pr (40)			74.8 La (300)	Ta (300)
	34.2 Tm(80)	Nd (50)		(50)Ba	73.2 Ta (200)	Cb (30)
	33.0 Ta (100)	Zr (35)			72.2 Ce (35)	Mn(20) Be (15)
(70)Fe	32.9 Pr (40)	Tb (30)		(150)Ti	71.9 Cr (50)	V (30)
	31.8 Os (100)	U (30)			70.0 Co (300)	Ir (50)
	30.8 Re (50)	Sm (40)			69.0 Rh (100)	Cr (50)
(70)Ti	29.3 Co (600)	Zn (35)			68.0 Ir (100)	Pr (30)
(40)Ba	28.7 Pr (200)	Co (125)			67.9 La (100)	Tm(35)
	27.4 Mo(80)	Mn(50)			66.8 Ta (100)	Co (100)Sm (100)
	26.4 Mo(100)	Cr (100)			65.5 Co (800)	Ta (200)
(100)Fe	25.7 Co (200)	Eu (50)			64.1 Co (35)	Tm(35)
	24.2 Mo(25)	V (20)		(100)Ti	63.7 Pr (100)	Nd (50)
(125)Ti	23.0 Co (150)	Mo(15)			62.3 Ce (40)	Ti (25)
	22.9 Ta (50)	Cr (30)			61.8 Tm(50)	Co (25)
	21.9 Cr (50)	Nd (40)			60.4 Sm (50)	Nd (50)
	20.8 Hf (50)	U (25)		(50)Ti	59.2 La (100)	Nd (40)
(100)Fe	19.5 Ta (300)	La (150)			58.4 La (100)	Mo(30)
	18.7 Fe (10)	Pr (10)			57.8 Ti (12)	Nd (10)
(200)Ti	17.2 Ru (12)	Mo(12)		(150)Fe	56.3 Ta (200)	Cr (40)
	16.1 Cr (300)	Ir (200)Os (150)		(125)Ti	55.3 Cs (2000)Tm(25)	
	15.9 Tm(200)	Sm (50) B(40)Band		(1000)Ba	54.5 Ru (1000)Sm (60)	
	14.0 Co (60)	Re (50)			53.6 Ta (200)	Co (25)
	13.3 Cr (150)	La (100)		(150)Ti	52.4 Sm (80)	Pt (60)
	12.0 Pr (60)	Dy (50) C(50)Band			51.9 Ta (400)	Os (150)
(200)Fe	11.2 Eu (50)	Sm (50)			50.4 Os (150)	Ir (80)
	10.8 Mo(10)	Pr (10)		(100)Fe	49.6 Co (600)	Ti (100)
	09.9 W (50)	Mo(40)		(125)Ti	48.7 Os (100)	Ir (100)
	08.0 Hf (25)	Rh (15)		(200)Fe	47.8 Ta (150)	Ni (30)
(50)Fe	07.3 Sr (1000)Mn(50)				46.9 Ni (50)	W (30)

Wave-length Table-chart

```
           45.9 Cr (200)   Ir (200)           (125)Fe  84.2 Os (100)   Co (60)
(150)Ti    44.6 Cr (100)   Mn(60)                      83.9 Co (100)   Gd (80)
           43.8 Co (500)   Sm (100)           (300)Fe  82.2 Ti (40)    Cr (25)
           42.4 Mn(80)     Nd (50)            (100)Ti  81.2 Tm(400)    Mg(100)
           41.2 Nd (50)    Cr (60)                     80.9Ta (200)    Cu (200)
           40.0 Gd (80)    Cr (80)            (70)Ti   79.8 Os (80)    Mn(60)
           39.6 Cu (100)   Os (100)                    78.4 Ir (200)   Sm (100)Co (100)
           38.4 Mn(40)     U  (25)                     77.2 Pr (125)   Re (40)
           37.8 Gd (150)   Os (50)            (500)Fe  76.0 Gd (100)   Yt (10)
           36.8 Mo(40)     Ti (40)                     75.3 Cr (40)    Re (30)
(80)Ti     35.7 Cr (125)   Pr (125)           (80)Ti   74.1 Gd (150)   Mo(125)
(100)Ti    34.1 Pr (150)   Mn(30)                      73.0 Sm (150)   Ru (100)
(150)Ti    33.9 Co (500)   Sm (40)                     72.7 Mn(100)    Sm (100)
           32.8 Ir (80)    Sm (60)            (100)Ti  71.2 Co (100)   Gd (50)
(125)Fe    31.1 Sr (10)    Pr (10)                     70.1 Mn(80)     Sm (60)
(200)Cu    30.9 Co (1000)Ta (300)Cr (150)     (200)Fe  69.5 Co (300)   V  (20)
           29.6 Os (80)    Tm(80)             (80)Ti   68.7 Pr (125)   Mo(25)
(600)Fe    28.6 Rh (500)   Ce (30)                     67.3 Sm (200)   Ba (20)  Re (75)
(100)Ti    27.4 Ta (150)   Co (100)Ce (50)    (500)Fe  66.5 Co (300)   Gd (200)
(100)Ca    26.9 Fu (100)   La (100)           (100)Ti  65.8 Pr (90)    Th (30)
(100)Fe    25.1 La (100)   Re (30)                     64.9 Eu (80)    Mn(60)
(80)Ba     24.7 Sn (500)   Os (80)                     63.4 Ce (35)    Ti (25)
(60)Ba     28.9 Sm (100)   Mn(50)  Ce (35)             62.4 Ni (150)   Mn(40)
(100)Ti    22.5 Eu (700)   Tm(200)Re (100)    (300)Fe  61.6 Ce (30)    Mn(30)
           21.0 Ta (200)   Gd (50)  La (200)           60.0 Ru (150)   Ce (60)
(40)Fe     20.0 Gd (50)    Pt (40)            (400)Fe  59.1 Ni (400)   Ta (30)
           19.6 Gd (150)   Sm (150)                    58.5 Sm (150)   Pr (90)
(130)Ti    18.5 Lu (300)   Er (40)            (150)Ti  57.4 Mo(50)     Zr (40)
(30)Fe     17.1 Co (300)   Ru (60)                     56.1 Sm (30)    Ca (20)
           16.8 Ru (100)   Re (80)            (150)Ti  55.3 Ca (100)   La (40)
           15.0 Sm (100)   Cr (25)            (200)Fe  54.4 Ca (200)   Re (100)Sm (100)
           14.5 Gd (60)    Co (60)            (150)Ti  53.3 Ti (80)    Mn(50)
           13.3 Re (300)   W  (30)                     52.7 Sm (200)   Tb (25)
(100)Ti    12.7 W  (30)    Mo(25)                      51.5 Mn(125)    Nd (100)
(200)Sn    11.3 In (5000)Ta (300)Sm (100)     (150)Ti  50.8 Ir (60)    Lu (40)
           10.9 Ta (200)   Pr (200)           (150)Ti  49.1 Pr (125)   Ru (125)
           09.3 Cu (150)   Ta (60)                     48.0 Tb (25)    U  (12)
           08.2 Fe (40)    Ru (15)            (200)Fe  47.7 Os (200)   Mn(60)
           07.5 Cu (50)    Re (40)                     46.3 Nd (100)   Gd (80)
           06.2 Gd (100)   Gd (80)                     45.7 Co (125)   Co (40)
(60)Ba     05.9 Yt (50)    Er (40)            (200)Fe  44.2 Sm (100)   Ru (40)
           04.8 W  (30)    Pr (20)            (200)Fe  43.1 Ti (80)    Mo(25)
           03.8 Mn(60)     Rh (30)            (400)Fe  42.5 Pt (800)   Mo(40)
           02.2 Mn(125)    Pr (10)                     41.6 Ta (100)   Ta (60)
(60)Ti     01.2 Nd (50)    Cr (40)            (80)Ti   40.3 Re (40)    Ce (20)
           00.2 Cr (50)    Er (20)                     39.7 Ru (125)   Yb (45)
           4400                                        38.1 Pr (50)    Gd (40)
           99.4 Sm (100)   Mo(20)                      37.2 Au (50)    Cb (40)
(100)Pt    98.8 Mn(150)   Ru (125)Gd (100)             36.3 Mn(80)     Os (80)
           97.1 Gd (150)   Na (70)            (100)Ca  35.5 Eu (2400)Fe (70)
(60)Ti     96.8 Cr (200)   Pr (200)Ta (100)   (150)Ca  34.3 Sm (200)   Ti (100)
           95.3 Ir (100)   Ti (20)            (150)Fe  33.8 Sm (200)   Fe (30)
(400)Fe    94.5 Co (100)   Na (60)                     32.3 Pr (80)    Cr (30)
(60)Ba     93.0 Tb (100)   Pr (25)            (60)Ba   31.8 Gd (60)    Sc (50)  Pr (60)
           92.1 Ir (35)    Cr (30)            (200)Fe  30.6 Gd (150)   Ti (35)
           91.6 Mn(50)     Mo(40)                      29.9 La (200)   Pr (200)
(40)Fe     90.0 Mn(100)    Fe (40)                     28.4 Ru (125)   V  (25)
(100)Fe    89.7 Ti (100)   Cr (25)            (500)Fe  27.3 Ti (125)   La (30)
(80)Ba     88.9 V  (60)    Os (60)            (80)Ti   26.2 Ir (400)   Gd (50)
           87.8 Pr (40)    Mo(25)             (100)Ca  25.4 W  (15)    Cr (15)
           86.9 Gd (100)   Co (50)                     24.3 Sm (300)   Pr (90)
(50)Fe     85.6 Pr (40)    Eu (40)                     23.6 Mo(40)     V  (40)
```

(300)Fe	22.5	Gd (100)	Ti (80)		61.2	Ru (40)	W (20)
(60)Ti	21.1	Sm (150)	Gd (100)Pr (100)		60.9	Gd (200)	Sm (100)
	20.4	Os (400)	Sm (200)	(100)Ni	59.9	Tm(300)	Cr (200)Pr (100)
	19.0	Gd (200)	Mn(100)Pr (100)	(70)Fe	58.3	Hg (3000)Re (80)
	18.7	Ce (40)	U (15)		57.5	Pr (25)	Re (15)
(80)Ti	17.2	Sm (80)	Eu (60)		56.8	Tb (60)	Hf (30)
(70)Ti	16.5	Tb (30)	Ce (25)	(50)Ca	55.0	Eu (150)	Ta (80)
(600)Fe	15.1	Sc (100)	Cu (40)		54.7	Eu (100)	La (80)
	14.8	Mn(150)	Gd (200)		53.1	Tb (50)	Mo (25)
	13.7	Pr (90)	Ce (35)	(300)Fe	52.7	Sm (125)	Ir (50)
	12.1	Pr (50)	Nd (40)		51.7	Cr (300)	Cr (100)
	11.8	Mn(100)	Gd (100)	(40)Ba	50.4	Sm (150)	Pr (70)
	10.0	Ru (150)	Mn(50)		49.7	Ce (40)	Ru (20)
	09.3	Sm (100)	Tb (40)		48.7	Yt (100)	W (50)
(125)Fe	08.4	Pr (125)	Gd (100)		47.4	Hg (200)	Sm (150)Pr (100)
(100)Fe	07.7	Ce (40)	Be (20)	(50)Fe	46.8	Cr (200)	Gd (150)Gd (100)
(20)Ba	06.6	Gd (70)	Re (60)		45.8	Sm (100)	Eu (80)
	05.8	Pr (100)	Ti (20)		44.5	Cr (400)	Pr (150)Gd (50)
(1000)Fe	04.7	Ti (50)	Pr (25)		43.9	Mn(100)	Cr (60)
	03.7	Ir (300)	Pr (100) Gd (100)		42.1	Gd (200)	Ru (60)
(80)Ba	02.4	Ta (100)	Os (50)		41.2	Gd (200)	V (60)
(60)Fe	01.5	Ni (1000)Gd (200)		40.1	Cr (80)	La (50)
	00.3	Sc (150)	V (60)		39.4	Cr (450)	Hg (150)
					38.6	Pr (100)	Tb (100)

4300

	99.4	Ir (400)	Ti (40)	(400)Fe	37.5	Cr (500)	Sr (150)Eu (100)
	98.0	Yt (150)	Ta (40)		36.1	Sm (100)	Tb (40)
	97.7	Ru (150)	Gd (100)		35.7	Pr (80)	Th (10)
	96.3	Ag (100)	Pr (80)		34.1	Sm (200)	La (40)
(80)Fe	95.2	V (60)	Ti (50)		33.7	La (800)	Pr (150)
	94.8	Os (150)	Re (100)	(20)Ba	32.5	Cr (125)	V (60)
(60)Ti	93.9	U (40)	Ce (35)		31.6	Ni (200)	Eu (80)
	92.5	Ir (100)	Re (100)Gd (100)		30.6	Gd (100)	Sm (80)
	91.3	Re (60)	Th (50)		29.0	Sm (300)	Eu (200)Gd (100)
(100)Fe	90.4	Ru (150)	Sm (150)Gd (100)		28.6	Os (60)	Pr (50)
	89.9	V (80)	Mn(50)	(100)Fe	27.1	Gd (500)	Pt (80)
(125)Fe	88.4	Mn(60)	Ti (25)	(60)Ti	26.4	Tb (150)	Mn(80) Tb (100)
(150)Fe	87.8	Eu (200)	Gd (150)	(1000)Fe	25.7	Gd (500)	Ti (100)Nd (100)
	86.4	Tm(200)	Ta (50)		24.0	Gd (100)	Sm (20) Cr (125)
	85.6	Ru (125)	Ru (125)		23.2	Sm (125)	Pr (100)Cr (100)
	84.9	Cr (150)	V (125)		22.5	La (150)	Eu (60)
1(000)Fe	83.5	Eu (100)	Gd (30)	(70)Ti	21.6	Cr (140)	Mn(60)
	82.6	Mn(80)	Ce (40)		20.5	Cr (125)	Sc (50)
	81.6	Mo(150)	Mn(80)		19.6	Cr (100)	Sr (50)
	80.6	Gd (100)	Pr (50)	(100)Ti	18.9	Sm (300)	Tb (150)
	79.2	V (200)	Pr (100)		17.8	Pr (40)	Pr (35)
	78.2	Cu (200)	Sm (100)		16.0	Gd (150)	Pr (50)
	77.0	Ir (110)	Cr (25)	(500)Fe	15.0	La (50)	Au (40)
	76.7	Cr (20)	Nd (15)	(100)Ti	14.8	Gd (100)	Sc (50)
(500)Fe	75.9	Ce (40)	Nd (30) Sm (200)		13.8	Gd (200)	Sm (40)
(150)Mn	74.8	Rh (1000)Yt (150)Sc (100)		12.5	Mn(100)	Ti (35)
(50)Fe	73.8	Gd (200)	Rh (60)		11.4	Ir (300)	Os (150)
	72.2	Ru (125)	Ir (40)		10.5	Ir (150)	Gd (100)
	71.2	Cr (200)	Pr (125)	(125)Fe	09.0	Sm (200)	Yt (50)
	70.4	Eu (60)	Os (50)		08.6	Dy (100)	Bi (50)
(200)Fe	69.7	Gd (250)	Gd (50)	(1000)Fe	07.9	Ti (100)	Ca (45)
	68.3	Pr (125)	Sm (60)		06.3	Gd (200)	U (40)
(160)Fe	67.5	Re (80)	Tb (30)	(300)Ti	05.9	Pr (150)	Fe (100)Cr (150)
	66.0	Pr (30)	Zr (25)		04.9	Sm (100)	Gd (100)
	65.6	Os (40)	Mn(25)		03.5	Pr (100)	Nd (100)
	64.6	La (50)	Ce (30)	(50)Fe	02.9	Ta (125)	Zr (100)W (60)
	63.4	Sm (60)	Pr (25)	(150)Ti	01.6	Ir (200)	Cr (100)
	62.0	Sm (150)	Sm (60)	(125)Ti	00.5	Cr (100)	Mn(60)

4200

(500)Fe 99.2 Cr (100) Ti (70)
(125)Ti 98.6 Fe (100) Ca (30)
        97.7 Cr (125) Gd (100)
        96.0 Gd (200) La (200)Sm (100)
(100)Ti 95.7 Cr (125) Ni (100)
(700)Fe 94.1 Ti (60) W (70)
        93.2 Mo(125) Os (60)
        92.1 Sm (100) Mo(100)
(125)Fe 91.4 Re (100) Ba (10)
(70)Ti 90.9 Zr (40) Fe (35)
(125)Ti 89.7 Cr (3000)Ce (50)
(50)Fe 88.7 Rh (400) Ni (150)Pt (75)
(100)Ti 87.4 U (15) V (15)
(100)Ti 86.9 La (400) Ir (200)Sm (100)
(125)Fe 85.4 Sm (200) Co (125)
        84.5 Mo(125) Mn(80)
(40)Ca 83.0 Sm (30) Ba (25)
(600)Fe 82.4 Pr (75) Ti (70) Th (25)
(80)Ti 81.0 Mn(100) Sm (50) Th (20)
        80.7 Sm (200) Gd (200)
        79.6 Sm (100) Sm (50)
(50)Ti 78.5 Tb (200) Pr (35)
        77.5 Lu (30) Mo(30) Th (20)
(50)Ti 76.7 Tb (50) Mo(30)
        75.1 Cu (80) La (40)
(100)Ti 74.8 Cr (4000)Gd (100)
        73.2 Li (200) Rh (25)
(40)Ti 72.2 Pr (50) Cr (40)
(1000)Fe 71.7 Fe (400) Ta (40)
(50)Ti 70.1 Cb (30) Ce (25)
        69.4 La (150) Cr (40)
        68.1 Ir (200) Ta (50)
(125)Fe 67.8 Gd (40) U (15)
(70)Fe 66.9 Gd (50) Cr (30)
        65.9 Mn(100) Ir (60)
        64.2 Fe (35) Ba (15) Ho (15)
(125)Ti 63.5 La (150) Cr (125)
        62.6 Sm (200) Gd (150)
(70)Ti 61.3 Cr (125) Cr (35)
(400)Fe 60.4 Os (200) Gd (30)
        59.1 Ir (200) Cr (35)
(70)Ti 58.5 Fe (60) Sm (50)
        57.5 Re (125) Mn(100)
(80)Ti 56.3 Sm (150) Zr (60)
        55.7 Ce (40) Cb (30)
        54.3 Cr (5000)Ho (100)
        53.6 Gd (10) Ce (40)
        52.3 Co (150) Cr (35)
        51.7 Gd (300) Sm (200)
(400)Fe 50.7 Fe (250) Pr (18)
(60)Ti 49.9 La (100) Sm (50)
(150)Fe 48.2 Cu (80) Ce (60)
(200)Fe 47.4 Pr (60) Nd (50)
(80)Fe 46.5 Gd (150) P (70) Sc (80)
(80)Ti 45.2 Ta (30) Dy (25)
        44.6 Sm (100) W (40)
        43.0 Ru (100) Gd (60)
        42.1 Tm(500) Cu (20)
        41.2 Zr (200) Ru (100)
        40.7 Cr (200) Zr (100)
        39.7 Mn(100) Zr (100)

(200)Fe 38.3 La (500) Gd (200)Cr (100)
        37.7 Cr (70) Sm (60)
        36.7 Sm (60) Ce (30)
(300)Fe 35.6 Eu (400) Mn(160)
        34.5 Sm (60) Cr (60)
(250)Fe 33.6 Fe (100) Co (100)
        32.5 Mo(125) Cr (70)
        31.7 Ce (30) U (25)
        30.9 La (150) Cr (70)
        29.1 Cb (50) Sm (40)
        28.6 Ta (25) U (18)
(300)Fe 27.4 Re (200) Zr (150)
(500)Ca 26.7 Ge (200) Cr (125)
(80)Fe 25.8 Gd (150) Pr (50)
(200)Fe 24.1 Fe (60) Cr (60)
        23.7 Sm (50) Ce (20)
(200)Fe 22.1 P (300) Pr (125)Cr (100)
        21.0 Re (100) Cr (80)
(80)Fe 20.6 Sm (100) Mn(60)
(250)Fe 19.3 Pr (30) W (25)
        18.0 Dy (50) Ir (20)
(200)Fe 17.5 La (200) Cr (150)Ru (100)
(200)Fe 16.1 Cr (60) Mo(10) Gd (100)
(60)Fe 15.5 Rb (1000)Sr (300)Gd (200)
        14.4 Ru (100) Cb (40)
(100)Fe 13.6 Cr (60) Zr (40)
(150)Ag 12.9 Pd (500) Ru (125)Gd (150)
        11.7 Dy (200) Os (150)Cr (100)
(300)Fe 10.3 Ag (200) Sm (50)
        09.3 Cr (100) Cr (80)
(100)Fe 08.6 Zr (100) Zr (30)
(80)Fe 07.1 W (25) Ru (20)
(125)Fe 06.7 Ru (100) Cr (80)
(50)Fe 05.0 Eu (200) Ta (100)
        04.0 La (200) Cr (80)
(200)Fe 03.7 Tm(250) Cr (100)
(400)Fe 02.0 Os (100) Ce (40)
        01.8 Rb (2000)Zr (50)
(80)Fe 00.9 Cr (80) Ni (40)

4100

(300)Fe 99.0 Ru (150) Tm(100)
(250)Fe 98.3 Cr (100) Ru (60)
        97.5 Ru (100) Cr (70)
(100)Fe 96.5 La (200) Rh (100)
(150)Fe 95.3 Os (100) Ni (30)
        94.9 Cr (70) Dy (50)
        93.6 Cr (100) Ce (35)
        92.4 Pt (100) La (50)
(200)Fe 91.4 Gd (100) Cr (50)
        90.7 Gd (100) Gd (100)
        89.5 Pr (100) Mn(80)
        88.3 Mo(100) Ti (35)
(250)Fe 87.6 Tm(300) Fe (200)
(100)Ti 86.8 Dy (100) Ce (80)
        85.8 Mo(40) Cr (30)
(100)Fe 84.2 Gd (150) Lu (100)
        83.3 Zr (40) Ir (40)
(80)Fe 82.9 Re (150) Ir (50)
(200)Fe 81.7 Ta (40) Mo(25)
(100)Ti 80.8 Er (25) Tb (20)
        79.4 Pr (200) Cr (100)
        78.9 Tb (50) Mo(25)

(100)Fe	77.5 Cu (60)	Yt (50)			15.7 Ir (100)	Ce (40)	
(100)Fe	76.5 Mn(100)	Ce (18)		(80)Fe	14.4 Mn(20)	Ru (20)	
(100)Fe	75.6 Os (100)	Ta (100)			13.3 Ru (40)	Mn(40)	
(100)Fe	74.9 Cr (100)	Yt (100)		(70)Fe	12.0 Os (150)	Ru (125)	
(50)Fe	73.2 Os (100)	Ho (50)			11.7 V (100)	Ce (35)	
(60)Fe	72.0 Ga (2000)	Ir (150)			10.5 Co (600)	Mn(80)	
	71.8 Pr (75)	Cr (70)		(120)Fe	09.8 V (40)	Cr (40)	
(80)Fe	70.9 Cr (70)	Gd (50)			08.6 Ho (100)	Cr (30)	
	69.8 Pd (200)	Cr (80)		(120)Fe	07.4 Ce (30)	Rh (25)	
	68.1 Cb (100)	Pb (20)			06.5 Sm (100)	Ce (30)	
	67.5 Ru (100)	Dy (50)			05.8 Tm(300)	Mn(50)	
	66.0 Ir (150)	Zr (50)		(100)Fe	04.1 Co (50)	La (40)	
	65.5 Cr (80)	Ce (40)			03.8 Ho (400)	Dy (50)	
	64.1 Pr (200)	Pt (100)			02.3 Yt (150)	Mn(100)	
	63.6 Cr (100)	Ho (100)			01.7 In (2000)	Fe (40)	
	62.7 Gd (50)	Mo(25)		(80)Fe	00.9 Cb (300)	Pr (200)	Ir (100)
	61.4 Cr (50)	Zr (40)			**4000**		
	60.2 La (40)	Pr (20)			99.5 La (100)	Cr (30)	
(60)Ti	59.6 Ce (30)	V (20)		(100)Fe	98.1 Gd (100)	Mo(20)	
(100)Fe	58.7 Os (50)	Co (25)			97.7 Ru (25)	Rh (25)	
(150)Fe	57.7 Mn(40)	Mo(25)			96.8 Pr (30)	Mo(20)	
(100)Fe	56.8 Zr (25)	U (15)		(80)Fe	95.9 V (40)	U (18)	
	55.7 Ir (80)	Mn(40)			94.1 Tm(300)	Pr (50)	
(200)Fe	54.8 Rh (60)	Lu (40)			93.1 Hf (25)	Ce (20)	
(120)Fe	53.9 Cr (50)	Cr (40)			92.3 Co (600)	Gd (100)	
(70)Fe	52.5 Cb (100)	Cr (50)			91.8 Os (100)	Ru (20)	
	51.9 La (200)	Ce (30)			90.4 Gd (100)	V (60)	
(50)Fe	50.2 Ti (35)	Zr (25)			89.9 Mn(80)	Yb (50)	
(100)Fe	49.3 Zr (100)	Re (40)			88.4 Os (100)	Co (50)	
	48.3 Ru (60)	Mu(50)		(50)Fe	87.3 Pd (500)	Gd (80)	Er (20)
(200)Fe	47.6 Ta (40)	Mn(40)			86.7 La (500)	Co (400)	
	46.7 Ru (100)	Dy (40)		(180)Fe	85.3 Gd (100)	Ru (40)	
	45.7 Ru (125)	Ti (15)		(120)Fe	84.4 Sm (80)	Mo(40)	
	44.1 Ru (150)	Re (125)			83.6 Mn(80)	Yt (50)	
(600)Fe	43.8 Pr (200)	Mo(100)		(60)Ti	82.7 Rh (100)	Mn(80)	
	42.8 Yt (100)	Ce (75)			81.2 Zr (150)	Pr (75)	
	41.7 La (200)	Pr (150)		(60)Fe	80.6 Ru (125)	Cu (30)	
	40.8 Pd (100)	Gd (20)		(80)Fe	79.7 Cb (500)	Mn(100)	
	39.7 Cb (50)	Fe (40)		(125)Ti	78.4 Fe (80)	Gd (20)	
	38.3 Tm(80)	Mo(20)		(400)Sr	77.3 La (600)	Hg (150)	Dy (150)
(100)Fe	37.8 Os (100)	Cb (100)		(80)Fe	76.6 Co (70)	Ru (60)	
	36.4 Re (150)	Ta (80)			75.8 Sm (40)	Cu (40)	
	35.2 Rh (300)	Os (200)		(80)Fe	74.7 Os (80)	W (50)	
(150)Fe	34.6 V (40)	Cr (25)		(80)Fe	73.7 Gd (100)	Dy (80)	
(50)Fe	33.4 Re (200)	Ce (35)			72.7 Zr (100)	Ce (20)	
(300)Fe	32.2 Li (400)	Gd (25)		(300)Fe	71.7 Ce (30)	Os (30)	
	31.4 Gd (100)	Mn(50)		(50)Fe	70.2 Mn(80)	Gd (80)	
(50)Ba	30.3 Gd (200)	Fe (20)			69.2 Th (40)	Ir (30)	
	29.3 Ta (200)	Eu (150)			68.5 Co (150)	Mn(50)	
	28.8 Rh (300)	Yt (150)		(150)Fe	67.9 La (150)	Ta (100)	
(100)Fe	27.1 Ho (150)	Ti (70)		(100)Fe	66.9 Os (100)	Tb (40)	
(80)Fe	26.5 Cr (100)	Cr (30)		(80)Ti	65.7 Cr (80)	Au (50)	
(80)Fe	25.6 Fe (25)	Ho (20)		(50)Ti	64.1 Zr (100)	La (40)	
	24.7 Lu (200)	Os (30)		(400)Fe	63.5 Mn(100)	Cu (30)	
(80)Fe	23.2 La (500)	Cb (200)		(120)Fe	62.6 Cu (500)	Pr (150)	Pb (20)
(70)Fe	22.5 Ti (40)	Cr (30)			61.7 Mn(80)	Ta (50)	
(100)Fe	21.3 Co (1000)	Rh (150)	Bi (125)	(60)Ti	60.3 La (80)	Tb (25)	
(80)Fe	20.2 Ho (50)	Cr (40)			59.8 Gd (50)	Ir (30)	
	19.6 Rh (100)	Er (18)		(80)Fe	58.9 Cb (1000)	Co (200)	Gd (100)
(200)Fe	18.7 Co (1000)	Pt (400)	Pr (250)		57.8 Pb (2000)	Co (100)	
	17.5 Ir (50)	Ce (30)			56.5 Pr (100)	Cr (30)	
	16.3 Rh (30)	U (25)		(80)Ti	55.2 Ag (800)	Zr (100)	

Wave-length Table-chart

```
              54.7 Gd (80)    Pr   (50)              (100)Ba  93.4 Cr  (60)     Ce   (50)
              53.9 Ho (400)   Gd  (200)                       92.8 Cr  (150)    Ir   (150)
              52.9 Co (40)    Tb   (40)                       91.1 Cr  (300)    Zr   (100)
              51.4 Ru (125)   Pr   (50)                       90.5 V   (125)    Co   (80)
              50.0 La (60)    Cu   (30)              (150)Ti  89.7 Pr  (200)    Cr   (80)
              49.9 Gd (100)   Gd   (80)                       88.5 La  (1000)V  (70)
              48.7 Cr (80)    Mn   (60)                       87.9 Yb  (1000)Er (100)Gd (100)
              47.2 K  (400)   Gd  (150)              (125)Fe  86.1 Mn  (40)     Pr   (40)
              46.5 Hg (200)   Cr   (30)              (125)Fe  85.3 Li  (100)    Mn   (75)
   (400)Fe    45.8 Co (400)   Ho (200)Dy (150)                84.3 Cr  (80)     Dy   (80)
    (70)Fe    44.1 K  (800)   Pr   (50)              (200)Fe  83.9 Cr  (200)    Dy (150)Sm (100)
              43.6 Cr (30)    Fe   (25)               (80)Ti  82.0 Pr  (125)    Yt   (60)
              42.9 La (400)   Ce   (50)              (150)Fe  81.9 Dy  (150)    Ti (100)Cr (100)
              41.9 Os (100)   Mn  (100)                       80.5 V   (40)     Ce   (35)
              40.8 Ho (150)   Ce   (70)                       79.5 Co  (150)    Gd   (100)
              39.1 Cr (100)   Pr   (50)                       78.5 Dy  (200)    Co   (100)
              38.0 Mo (20)    Mn   (15)              (300)Fe  77.7 Os  (300)    Mn   (50)
              37.9 Gd (200)   Os   (80)                       76.6 Cr  (300)    Tb   (150)
              36.1 Eu (50)    Th   (15)                       75.2 Zr  (50)     Sc   (50)
    (50)Ti    35.5 Co (150)   Mn   (50)                       74.7 Co  (100)    Gd   (100)
              34.4 Mn (250)   Ti   (25)              (200)Ca  73.5 Ni  (800)    Co   (150)
   (100)Ta    33.0 Mn (400)   Tb (125)Ir (100)                72.1 Pr  (125)    Co (100)Ni (100)
    (80)Fe    32.9 Ga (1000)  Ti   (35)              (200)Fe  71.9 Eu  (1000)Pr (100)
    (80)Fe    31.6 La (400)   Tb   (50)               (50)Fe  70.0 Ta  (100)    Ni (40)Cr (200)
   (120)Fe    30.7 Mn (500)   Ti   (80)              (600)Fe  69.2 Gd  (200)    Os (100)Co (100)
    (80)Fe    29.6 Re (80)    Ta   (50)              (500)Ca  68.4 Dy  (300)    Ag (100)Zr (100)
              28.9 Zr (40)    Ce   (35)              (125)Fe  67.4 Fe  (60)     Pr   (40)
              27.0 Co (200)   Zr  (100)              (180)Fe  66.5 Pr  (100)    Pt   (80)
    (70)Ti    26.1 Cr (100)   Mn   (50)                       65.2 Pr  (100)    Eu   (15)
              25.0 Cr (100)   La   (50)               (80)Fe  64.8 Pr  (125)    Pr (60) Ti (80)
   (120)Fe    24.7 Ti (80)    Tb   (40)              (125)Fe  63.6 Os  (500)    Cr   (300)
              23.4 Co (200)   Sc  (100)               (80)Ti  62.4 Re  (100)    Pr   (60)
              22.6 Cu (400)   Cr   (80)                       61.5 Al  (3000)Zr (500)Os (125)
   (200)Fe    21.8 Ti (100)   Mo   (50)                       60.9 Co  (60)     Pr   (50)
              20.9 Co (500)   Ir (80) Sc (50)                 59.5 Sm  (50)     Pr   (25)
              19.3 Co (80)    Tb (40) Th (8)         (150)Ti  58.6 Pd  (500)    Zr (500)Rh (200)
    (50)Fe    18.1 Mn (80)    Os   (60)               (80)Ca  57.6 Gd  (300)    Co (100)Tm(200)
    (80)Fe    17.1 Ti (70)    U    (25)              (250)Fe  56.6 Ti  (100)    Ce   (30)
              16.2 Ti (30)    Pr   (25)                       55.7 Eu  (50)     Fe   (25)
    (70)Ti    15.3 La (100)   Pr   (50)                       54.5 Dy  (40)     Nd   (40)
   (200)Fe    14.5 Ce (60)    Cr   (40)               (80)Fe  53.5 Pr  (150)    Gd   (100)
   (280)Fe    13.9 Co (300)   Ti   (70)               (80)Fe  52.9 Co  (100)    Gd   (100)
              12.2 Nd (80)    Ce   (60)              (150)Fe  51.1 Cr  (50)     Mn   (40)
              11.5 Re (35)    Mo   (25)                       50.3 Yt  (60)     Dy   (50)
              10.4 Nd (20)    Eu   (20)              (150)Fe  49.1 La  (1000)Pr (150)
   (120)Fe    09.7 Ti (60)    Gd   (50)              (150)Fe  48.7 Fe  (125)    Ti   (80)
    (80)Ti    08.7 Pr (150)   Ti (50) W (45)         (120)Fe  47.6 Pr  (125)    Ti   (70)
    (80)Fe    07.2 Sm (50)    Er   (35)                       46.8 Tb  (150)    Ir   (50)
    (60)Fe    06.3 Ta (30)    Ru   (25)                       45.5 Gd  (200)    Co   (200)
   (250)Fe    05.2 Tb (100)   Ti   (35)                       44.0 Al  (2000)Dy (300)
              04.4 Os (50)    Re   (30)                       43.6 V   (50)     Eu   (50)
    (50)Ti    03.8 Os (50)    Ce   (40)              (100)Fe  42.4 Mn  (75)     Gd   (60)
              02.5 Tb (50)    Ti   (40)               (60)Fe  41.7 Co  (200)    Cr   (200)
    (80)Fe    01.4 Cr (200)   Gd   (80)              (150)Fe  40.8 Co  (100)    Pr   (80)
              00.4 Dy (400)   Pr   (50)                       39.6 Tb  (300)    Os   (50)
                         3900                                 38.5 Os  (125)    Cr   (40)
              99.2 Ce (80)    Pr   (50)               (80)Fe  37.3 V   (50)     Pr   (50)
   (150)Ti    98.6 Fe (150)   V   (100)                       36.2 La  (100)    Re   (40)
   (300)Fe    97.3 Co (200)   Pr (100)Dy (200)       (100)Fe  35.9 Co  (400)    Pr   (125)
              96.6 Tm (200)   Ta  (100)              (100)V   34.8 Ir  (200)    Gd (100)Rh (100)
    (60)Fe    95.3 Co (1000)  La (600)Tm (340)       (600)Ca  33.6 Fe  (200)    Sm   (200)
              94.8 Pr (300)   Nd   (80)               (80)Fe  32.6 U   (35)     Dy   (30)
```

	31.5	Dy (200)	Os (40)			15.0 Li (200)	Ir (150)	Cr (125)
(600)Fe	30.5	Eu (1000)	Eu (100)		(50)Ti	14.3 Zr (70)	Dy (50)	
	29.2	La (400)	Re (100)	Zr (100)	(100)Fe	13.6 Pr (80)	Ti (40)	
	28.6	Cr (150)	Sm (60)			12.8 Pr (150)	Ce (50)	
(500)Fe	27.9	Pr (80)	Nd (80)			11.8 Sc (150)	Ti (40)	
	26.4	Mn (40)	Cr (35)			10.7 V (35)	Fe (30)	
(130)Fe	25.4	Tb (150)	Pr (125)		(50)Ba	09.9 Co (200)	V-50	
(70)Ti	24.1	Pr (100)	V (35)			08.7 Cr (200)	Pr (200)	
	23.4	Ru (60)	Ir (30)		(100)Fe	07.1 Eu (1000)	Sc (125)	Gd (100)
(600)Fe	22.9	Pt (100)	Ta (200)	Co (100)	(300)Fe	06.4 Co (150)	V (50)	Er (25)
	21.5	La (400)	Cr (150)	Zr (120)		05.6 Gd (50)	Nd (40)	
(500)Fe	20.2	Re (40)	V (35)		(70)Ti	04.7 V (20)	Th (20)	
	19.1	Cr (300)	Tb (40)		(100)Fe	03.9 Sm (60)	Cr (35)	
	18.8	Pr (100)	Ce (60)		(500)Fe	02.9 Mo (1000)	Cr (100)	Gd (100)
(150)Fe	17.1	Re (100)	Co (80)			01.7 Os (150)	Ru (50)	
(100)Fe	16.0	La (400)	Gd (150)	Cr (100)	(60)Fe	00.5 Zr (100)	Tm (80)	

Appendix

CONVERSION TABLE

1 angstrom (Å) = one ten-millionth of a millimeter
1 micron (μ) = 10,000 angstroms = 0.001 mm
1 millimeter (mm) = 0.039 in.
1 centimeter (cm) = 10 mm = 0.39 in.
1 meter (m) = 100 cm = 39.37 in. = 1.09 yd
1 kilometer (km) = 1000 m = 0.62 mi
1 milligram (mg) = the thousandth part of a gram
1 centigram (cg) = 10 mg
1 decigram (dg) = 10 cg = 1.54 gr
1 gram (g) = 1000 mg = 15 gr = 0.035 oz avoir
1 kilogram (kg) = 1000 g = 2.2 lb avoir
1 ounce (oz) = 437 gr = 28.35 g
1 pound (lb) = 16 oz = 453.59 g
1 ton = 2000 lb (avoirdupois)
1 liter (l) = 0.908 qt
1 pint (pt) = 1 pound approximately = ½ qt
1 horsepower-hour = (hp hr) = 745.7 watt-hours
1 kilowatt = 1000 watts
Velocity of light = 186,284 mi per sec

The wave-length of a spectral line may be found by dividing the speed of light by the frequency of the vibrations. The wave-number is the number of waves per centimeter; it is always more than 10 times the wave-length. Multiplying amperes by ohms gives volts. Multiplying amperes by volts gives watts. The focal length of a concave mirror is half its radius of curvature. What is termed the focal length of a concave grating is equal to the full length of its radius of curvature, and represents the distance between slit and grating in the spectroscope.

BIBLIOGRAPHY

Data of Geochemistry, Washington, D. C., Government Printing Office, 1924. F. W. Clarke.
Wavelength Tables, New York, John Wiley & Sons, Inc., 1939, G. R. Harrison.
Technical Analysis of Ores, New York, Chemical Publishing Co., Inc., 1939, F. G. Hills.
Dana's Textbook of Mineralogy, New York, John Wiley & Sons, 1932, W. E. Ford.
New Modern Encyclopedia, New York, Wm. H. Wise & Co., 1946.
Smith's Inorganic Chemistry, New York, Century Co., 1926, James Kendall.
Microscopic Determination of Minerals, Washington, D. C., Government Printing Office, 1934, Larsen & Berman.
Atomic Energy, Princeton, N. J., Princeton University Press, 1946, H. D. Smyth.
Handbook of Chemistry & Physics, Cleveland, Ohio, Chemical Rubber Publishing Co., 1935, C. D. Hadyman.
Chemical Spectroscopy, New York, John Wiley & Sons, Inc., 1939, W. R. Brode.
Spectroscopy, London, Blackie & Son Limited, 1933, Judd Lewis.
Common Minerals & Rocks, Boston, Mass., D. C. Heath & Co., 1907, W. O. Crosby.
Proceedings of the Summer Conferences, Cambridge, Mass., Technology Press, 1939.
Manual of Clinical Therapeutics, Philadelphia, Pa., W. B. Saunders Co., 1943, W. C. Cutting, M.D.
Field Book of Common Rocks and Minerals, New York, G. P. Putnam's Sons, 1931, F. B. Loomis.
Optics & Service Instruments, Brooklyn, N. Y., Chemical Publishing Co., 1941.

Firms Selling Spectroscopic Materials

Aluminum Co. of America, New Kensington, Pa. (chemicals)
American Lens Co., 313 Vine Ave., Park Ridge, Ill. (lenses)
Applied Research Lab., Glendale 4, Calif. (industrial spectrographs)
Baker Chemical Co., Phillipsburg, N. J. (chemicals)
Bausch and Lomb Optical Co., Rochester, N. Y. (lenses, prisms, instruments)
Central Scientific Co., Chicago, Ill. (replica gratings)
Cutting Sons, Campbell, Calif. (Plane gratings, spectroscopes)
Fairmont Chem. Co., 126 Liberty St., New York, N. Y. (rare metals)
Gaertner Scientific Corp., Chicago, Ill. (spectrometers)
Jarrell-Ash Co., 165 Newbury St., Boston, Mass. (chemicals)
John Wiley and Sons, New York, N. Y. (complete spectral tables)
Linde Air Products, 30 E. 42nd St., New York 17, N. Y. (rare gases)
Mackay, 198 Broadway, New York, N. Y. (rare metals)
Mechanical magazines, classified ads. (cheap chemicals)
Merck and Co., Inc., Rahway, N. J. (chemicals)
National Carbon Co., Cleveland, Ohio (carbon electrodes)
Spencer Lens Co., Buffalo, N. Y. (spectroscopes, lenses)
Welch Scientific Co., 1515 Sedgewick St., Chicago, Ill. (gratings)

INDEX

A

Absorption spectra, 58
Air spectrum, 17
Alkali metals, 153
Alpha particle, 150
Alumina, 91
Amyl acetate, 54
Analysis, 78, 83, 87
  automatic, 32
  by densitometer, 38
  by manganese, 84
  by paired lines, 83
  by standards, 76
  of hydrocarbons, 33
  precision method, 85
  qualitative, 58, 62
  quantitative, 71, 76
Ångstrom unit, 59, 213
Arc, 7, 8
  carbon, 47, 9
  construction, 7, 12
  currents (table), 8
  electric circuit, 11
  high-voltage, 12
  images of, 14, 21
  limitations, 14
  lines (tables), 174
  manipulations, 13-15
  portable, 8
  precautions, 12, 16
  temperature, 8
  voltage, 8
Atom in spectroscopy, 3
Atomic structure, 3

B

Balmer formula, 2
Band spectra, 59
Blowpipe, 88

Bohr's theory, 2, 3, 150
Bunsen, 2, 109, 153
Bunsen flame, 7

C

Cadmium unit, 105
Calcareous minerals, 93
Calcium fluoride bands, 59, 116
Carbon arc, 9
Carbonates, 94
Cobalt glass, 58
Collimator, 21
Color range (table), 61
Comparator, 34
Composition of the earth, 89
Compounds (table), 76
Condenser for spark, 17
Condensing lens, 21
Crookes, 163
Crystal forms, 96
Curie, 147
Cyanogen bands, 59, 107, 156

D

Densitometer, 36
Density of lines, 85
Diffraction, 19
Discharge tubes, 18
Dispersion, 24, 25
  concave grating, 24
Dolomite, 93
Doublets (table), 69
Dust lines, 59

E

Earth composition, 89
Einstein, 150
Electrodes, 8, 10, 31

Electron levels, 5, 6
  shells, 4, 71
Electrons, 3, 70
Element identification, 62
  symbols, properties, 73-75
Elements, 3
  characteristic lines, 80-173
  in rocks (table), 90
  periodic chart of, 3, 4

F

Film holders, 34, 55
Fluorescence, 117, 173
Focal lengths, 213
Fraunhofer grating, 2
  lines, 2, 58

G

Gamma radiations, 150
Gas-tube spectra, 17
Geiger-Muller counter, 98, 168
Geissler tube, 7, 18, 124, 159
Gems, 99, 111
Gold spectrum, 120
Graphite electrodes, 31
Gratings, diffraction, 2, 19
  concave, 2, 23, 47
  concave replicas, 54
  replicas, 23

H

Halogens, 4, 127
Hardness scale, 96
Huygens wave theory, 1
Hydrogen spectrum, 5

I

Identification of unknowns, 63
Induction coil, 17
Inert gases (table), 129
Infrared spectrum, 1, 33
Intense lines, 116, 132, 148
Intensities, 65
Intensity table, 80, 84

Ionization, 15, 171
Ions, 34
Iron spectrum, 129

K

Key lines, 66-68
Kick-back, 17
Kirchhoff, 2, 109, 153

L

Lengths and weights (table), 213
Lens, condensing, 53
  correction, 29
  distortions, 22
  power of (table), 28-29
Lenses, 22
  achromatic, 22
  focal length, 21
Light, 4
  speed of, 5, 213
Light-sources, arc, 7, 14
  spark, 15, 18
Line series, 5-6
Logarithmic sector, 86

M

Magnesian minerals, 93
Masks, 44, 53
Maxwell's theory, 2
Meggar's spectrographs, 2
Mendeleeff, 119, 122
Michelson's instruments, 2
Micron, 213
Mineral classification, 91
  species, 3
Mineralogy, 88
Molecular weights (table), 79
Multiple standards, 78
Multisource unit, 34-35

N

Negatron, 3
Neon tubes, 139

Index

Neutrino, 3
Newton's prism, 1
Nonmetallic minerals, 95

O

Oculars, 42, 45, 49
Ore manipulations, 15
Oxides, 91, 92

P

Periodic chart of elements, 3, 4
Permanganate of potassium, 58
Persistent lines, 64, 159
Planck, quantum theory, 2
Platinum group (table), 144
Positrons, 3
Potash minerals, 94
Preparation of samples, 18
Prices of instruments, 38, 46
Prisms, glass, 20, 41
 for infrared spectrum, 33
Protons, 3

Q

Quadruplet lines, 70
Qualitative analysis, 58, 62
Quanta of light, 4
Quantitative analysis, 71, 76
Quantometer, 32
Quartering the sample, 88
Quartz diluent, 15
Quintuplets, 70

R

Radioactive elements, 98, 147, 149, 150, 163, 168
Rapid testing, 84, 88
Rare earths (table), 115
Replica gratings, 23, 54
Resistors, 8-9
Resolution, 24
Rock composition, 89-90
Rowland's concave grating, 2

S

Scale, arbitrary, 28
 comparator, 37
 for prism instrument, 43
 of hardness, 96
 plotting, 131, 132
 spectrum, 22
 use of, 64
Silica as a diluent, 76
Silicates, 92, 93, 157
Singlet lines, 69, 70
Slits, 20, 43, 48
Sodium doublet, 158
 minerals, 94
Solar spectrum, 1, 2, 58, 122
Spark circuit, 17
 gap, 7
 manipulations, 17-18
 precautions, 16
 spectra, 15
 voltage, 7
Spectra, absorption, 58, 165
 bright-line, 59
 comparison, 63
 complex, 68, 168
 continuous, 58
 molecular, 59
 of alkalies, 70
 of gases, 7, 17, 18
 of metals, 7
 of nonmetals, 7, 18
 of rare earths, 115
 of solutions, 18
 of stars, 2
 orders of, 54
 ultraviolet, 23
 vanishing, 136
 weak, 145, 163
Spectral series, 2
 types, 68
Spectrogram, 83
 with sector, 86
Spectrograph, 22
 industrial, 29-30
 infrared, 33
 large, 25
 mass type, 33

Spectrograph—*Continued*
  operation of, 31
  plane grating type, 44-45
  quartz type, 25
  semicircular type, 56
Spectrometer, 46
  callibration, 51
  concave grating type, 46
  operation, 50
  scale for, 49-50
  scale symbols for, 52
Spectroscope, box type, 40
  construction, 38
  direct vision type, 20
  function, 19
  optical system, 20, 22
  plane grating type, 39
  prism type, 25-26
  telescope for, 21, 25
  types, 19, 40
Spectrum, dark line, 58
  diagrams of, 63
  intensities of, 72
  reversal of, 72
  suppression of, 85
  weak, 73
Spinthariscope, 150
Standard powders, 73
Subatomic particles, 3

T

Tesla coil, 18
Titanium minerals, 94

Trace elements, 95
Transformer for spark, 16
Triplet lines, 70

U

Ultraviolet spectrum, 1, 23, 31
Uranium nitrate spectrum, 58

V

Vaporization of chemicals, 15
  of crystals, 14
  of liquids, 15
  of ores, 14

W

Wavelength, 5, 213
  graphs, 26, 27
  series, 5
  table, 186-212
  tables, 61
Wavenumber, 213
Wollaston's spectroscope, 1

X

X-ray spectra, 121, 124, 136

Y

Yttrium bands, 171